Fundamentals of Planning and Designing Sustainable Post-Disaster Shelters

Avi Friedman • Baichi Chaki

Fundamentals of Planning and Designing Sustainable Post-Disaster Shelters

Springer

Avi Friedman
School of Architecture
McGill University
Montreal, QC, Canada

Baichi Chaki
Stantec Architecture Ltd.
Edmonton, AB, Canada

ISBN 978-3-031-83316-8 ISBN 978-3-031-83317-5 (eBook)
https://doi.org/10.1007/978-3-031-83317-5

This Springer imprint is published by the registered company Springer Nature Switzerland AG
The registered company address is: Gewerbestrasse 11, 6330 Cham, Switzerland

Preface

Every year, millions of people are forced to leave their homes due to war, environmental disasters, or persecution. According to the United Nations High Commissioner for Refugees (UNHCR), in 2022, around 108.4 million were forcibly displaced worldwide due to persecution, conflict, violence, or human rights violations. In most cases, the refugees start living permanently in the host region or countries in a long-lasting crisis. A study revealed that, at the end of 2017, the average duration of a refugee's exile was ten years. Not only following a war crisis and after a natural disaster such as earthquake, cyclone, flood, and riverbank erosion, it is essential to build many durable shelters rapidly.

The presence of refugees in the host region or country can cause many negative impacts on the environment, including deforestation, erosion of fertile land, and pollution of water sources. These complex challenges require camp planning, political will, and coordinated action to address the pragmatic aspects of immediate solutions. Current planning approaches for dwellings and neighborhoods for the displaced people face severe challenges during the expansion of the camp, building long-lasting shelters, and a need to provide amenities. The amenities necessary for their sustainable living conditions are clinics or health care, schools, places of worship, community gathering areas, open spaces, markets, and commercial areas.

In today's growing numbers of displaced people and refugee crises worldwide, designers must look for creative and innovative ways to solve these enormous housing challenges. The need to think innovatively about emergency housing led to the idea to write this book. The intention is to offer information on contemporary emergency shelter and planning concepts and illustrate them with outstanding international examples. This book is meticulously crafted to address a diverse readership ranging from architects, urban planners, and humanitarian aid workers to policymakers, researchers, and students in relevant fields. With a clear focus on the pressing global issue of displaced populations, this book offers a comprehensive and accessible guide to creating shelter solutions that prioritize functionality, efficiency, and human dignity. It serves as an essential resource for professionals seeking practical insights into designing shelters that cater to the unique needs of refugees, taking into consideration cultural, social, and environmental factors. Additionally, the book's thoughtful analysis and case studies make it an indispensable reference for academics and decision-makers invested in fostering innovative and sustainable approaches to refugee shelter design on an international scale.

After a disaster occurs, preliminary efforts focus on rescuing lives, followed by sheltering victims. These shelters are often constructed primarily by refugees who are most likely untrained in construction. The spontaneous construction without considering future needs often leads to unusable and unhealthy living conditions. Chapter 1 introduces the need for emergency shelters, the terminologies used in refugee housing, common approaches by the shelter organization, and long-term concerns associated with the shelter designs.

The critical elements that influence the selection of a suitable site for different types of settlements for refugees are explored in Chap. 2. The importance of site size and area per capita is emphasized at the outset of the chapter, along with the connection between health issues and overpopulation. The benefits of situating camps close to existing facilities are highlighted, emphasizing the need to avoid overburdening local resources and to promote integration. A thorough recommendation is provided for design consideration regarding the site's natural conditions, which include things like slope, soil quality, groundwater, flood plains, flora, wildlife, and microclimate.

In Chap. 3, the necessary considerations for a long-term stay in refugee accommodation are described, together with standards for sustainable site layout. Given their substantial influence on the ecological footprint and carbon emissions of housing, this chapter highlights the need to apply macro thinking to the design and density of housing. The need for contingency planning is emphasized throughout the chapter in order to gather environmental and demographic data that will help identify hazards and guide site development. To improve sustainable refugee living situations, a mixed-use modular site design is recommended. The density issues are discussed where low, medium, and high-density categories are examined. Each has unique benefits and problems. Environmental factors, which take into account things like geography, grading, and nearby resources, emphasize the need to combine site design with environmental protection.

Chapter 4 discusses the critical significance of emergency preparedness and response in delivering refugee shelters rapidly. It emphasizes the need for thorough needs assessments, particularly focusing on the specific requirements of women and children. It stresses the importance of establishing a multi-functional team for comprehensive coordination. The chapter highlights the necessity of contingency planning through examples such as the proactive measures taken in Cambodia and the response to the Mosul crisis. It also delves into the use of computational and parametric design, showcasing the Hex House as an innovative, scalable, and easily deployable solution for refugees, providing a comprehensive understanding of the entire process from emergency to long-term living solutions.

The infrastructure and utilities required to keep refugee camps safe and sustainable in times of emergency are the main topics of Chap. 5. The drainage system is essential, and careful design is needed to guarantee effective water management, reduce erosion, and stop floods. Designers should avoid regions that are prone to flooding because inadequate drainage might harm access points and shelters. Sanitary sewers, solid waste management, wastewater treatment, and pest control are vital aspects of maintaining hygiene and preventing diseases within the camp. One of the fundamental human rights is access to clean water, which is made possible by extensive water supply networks. The chapter also emphasizes how important it is to provide communal spaces like markets, schools, and houses of worship to foster involvement and well-being within the community. A case study of Rohingya refugee camps in Bangladesh shows how an entire settlement strategy may efficiently fulfil infrastructure and utility demands while reducing environmental impact.

The refugee shelter design requirements are introduced in Chap. 6. Techniques covered here range from conceptual approaches to living areas, access, space design, structural issues, and human comfort. The crucial relationship between shelter layout and function is also covered in this chapter, with particular attention to various types of shelters that are specifically designed to meet the requirements of refugees. Refugee shelters are categorized into three groups—emergency, transitional, and durable shelters—depending on whether they provide short-term or long-term housing options. In the months that follow wars or natural catastrophes, emergency shelters prioritize safety and privacy while meeting urgent housing demands. By allowing for progressive modification and component reuse, transitional shelters serve as a link between emergency situations and longer-term solutions. Long-term housing is provided through durable shelters, which include contextual, climatic, and cultural factors. Along with aspects of interior environmental quality, including natural lighting, ventilation, thermal comfort, personal security, and fire safety, the chapter also highlights variables like wind, flood, and earthquake resistance. A case study by the Balbek Bureau showcases creative modular town systems that provide internally displaced Ukrainians with respectable places to live. This illustrates how shelter form and function may be effectively combined to meet humanitarian issues.

Chapter 7 covers a wide range of prefabrication techniques, including benefits, drawbacks, and case studies. Prefabricated shelters, which are built in a factory and placed on location, frequently have walls composed of polyvinyl or glass, steel or aluminum frames, and roofs. They provide advantages such as lower worker costs, improved craftsmanship, weather protection during manufacture, and quicker construction. The shelters are appropriate for a variety of settings and can be utilized as temporary or permanent housing. These shelters are made using a variety of materials and techniques, including modular, panelized, and "kit of parts" approaches. The usage of prefabricated shelters in disaster-stricken locations is demonstrated through a case study pertaining to the Tohoku Earthquake in Japan. The case study covers many aspects of construction, obstacles, and site layouts.

One of the most important issues for planners and architects regarding the rising frequency of natural disasters and civil wars is the endeavor to develop a rational and prompt response to emergency shelters. The concept of self-help housing, its different forms (autonomous, organized, and employed), and its advantages—such as cost savings and resident empowerment—are covered in Chap. 8. It also explores the development, history, and use of self-help housing for refugees since the 1920s. The involvement of UNHCR and other organizations is highlighted while discussing the possibility of self-help housing as a temporary solution for refugees. There are guidelines for planning and building self-help shelters, including the kind of structure to use, what materials to use, and certain work procedures. The chapter includes a case study highlighting the Liina transitional shelter, a sustainable and flexible self-help housing option, and emphasizes the value of building training and quality assurance. In summary, this chapter highlights the value of self-help housing in solving the pressing need for shelters for refugees while fostering self-sufficiency and community reconstruction.

Chapter 9 focuses on different building materials, shows how to assess their sustainability, and talks about both traditional materials and some of the greenest options out there. The beginning of the chapter emphasizes the astonishing rates of depletion of natural resources and their negative impacts on the ecosystem, including carbon emissions and habitat destruction. It highlights the significance it is to control resource usage to protect ecosystems, using measures like the ecological footprint. The chapter continues by outlining the need to carefully choose building materials and account for their embodied energy values. Natural wood, bamboo, adobe bricks, clay or burnt bricks, wattle and daub reinforced mud walls, corrugated galvanized iron, plastic sheets, thatching materials, cement, cement blocks, and quarry materials are among the materials that are frequently utilized in the construction of refugee shelters. The environmental impact of each material is evaluated, emphasizing the necessity for responsible sourcing, production, and disposal procedures. The chapter concludes with a case study of Shigeru Ban's inventive Paper Log Houses, which represent affordable, recyclable, and simple-to-assemble shelters that have been used to assist survivors of disasters all around the world.

The topics covered in each of the previous chapters are essential to the design of long-term emergency shelters and communities. Ideally, planners and designers should incorporate all of them. Compromises are often necessary because refugee housing communities in many nations do not offer all the amenities and shelters that adhere to sustainable values. Even though displaced persons can spend years living in these settlements, they are denied basic human needs. This final chapter covers the practical aspects of building long-term emergency shelters. It offers an outlook on the design and construction of refugee shelters, specifies basic processes planners might follow at the commencement of their design, and suggests solutions for these escalating situations.

The book's trust is to offer a comprehensive planning and design knowledge to initiators of post-disaster shelters be them policy makers, representatives of aid agencies and professionals. We did our best to address other subjects associated with post-disaster shelters such as social and psychological factors, but issues of technical nature were central to our main objective. The chapters have been organized to follow the design process and be cumulative in the contemporary information that they offer.

Finally, we might anticipate a significant migration and displacement as temperatures rise and extreme weather events are more common and severe. Future years will only bring more suffering, upheaval, and instability if meaningful action is not taken now. To mitigate climate change and increase the resilience of vulnerable communities, everyone worldwide needs to work together to develop feasible approaches and promote sustainability.

Montreal, QC, Canada Avi Friedman
Edmonton, AB, Canada Baichi Chaki

Acknowledgments

The subjects of housing and shelters' design were central to our work for long and involved exchange of ideas with many colleagues. We would like to thank those who have directly or indirectly contributed to this book. Our apology if we have mistakenly omitted the name of someone. I will do my best to correct such omissions in future editions.

We want to express our sincere gratitude to our families, whose constant encouragement and support served as the cornerstone for this endeavor. To Baichi Chaki's parents: thank you for your unending love, guidance, and sacrifices; they have influenced her path and inspired her goals. Baichi Chaki's husband has been an inspiration and a source of strength, unwavering understanding, patience, and belief in her aspirations. To Natalie Qiu for helping with background research and to Kaite Fairbrother for proofreading the text. Our gratitude to Andrea Brody-Barre, our editor at Springer Nature, and to Nivetha Moorthi, Production Editor, for their guidance and support.

Avi Friedman
Baichi Chaki

Acknowledgement

Contents

1 The Global Need for Shelters 1

2 Site Selection 21

3 Site Planning 39

4 Planning for Rapid Delivery 59

5 Infrastructure and Utilities 77

6 Form and Function of a Shelter 97

7 Prefabricated Shelters 115

8 Designing Shelters for Self-Help Building 131

9 Sustainable Materials for Shelters 147

10 Charting a Path Forward 167

Weblink of Case Studies 179

Bibliography 181

Index 189

Abbreviations

BSRP	Building Safety and Resilience in the Pacific
DRR	Disaster risk reduction
EPS	Expanded polystyrene
GDP	Gross domestic product
GIS	Geographic information system
GSC	Global shelter cluster
IAQ	Indoor air quality
ICF	Insulated concrete forms
IDP	Internally displaced persons
IEQ	Indoor environmental quality
IOM	International Organization for Migration
NFI	Non-food item
UASC	Unaccompanied Asylum-Seeking Children
UNHCR	United Nations High Commissioner for Refugees
WASH	Water, sanitation, and hygiene
WCED	World Commission on Environment and Development

Illustration Credits

Figures not listed below are in the public domain or have been conceived, drawn or photographed by the authors. Every effort has been made to list all contributors and sources. In case of omission, the author and the publisher will include appropriate acknowledgment or correction in any subsequent edition of this book.

Fig. 1.1	After Alia Chughtai/Al Jazeera (2020). Year-end figures of displacement over the last decade
Fig. 1.2	Photo by Ahmed Akacha on Free Photo Domain (Pexels)
Fig. 1.3	After data by Ezinna E Enwereji (2009). Assessing Interventions Available to Internally Displaced Persons in Abia State, Nigeria
Fig. 1.4	After UNHCR (2019). Refugee returns by decade
Fig. 1.5	Photo by Ahmed Akacha on Free Photo Domain (Pexels)
Fig. 1.6	Photo by Ahmed Akacha on Free Photo Domain (Pexels)
Fig. 1.8	Photo by Ahmed Akacha on Free Photo Domain (Pexels)
Fig. 1.10	After Clement et el. (2021). Groundswell Part 2: Acting on Internal Climate Migration
Fig. 2.1	After Chiara and Koppelman (1975). Urban planning and design criteria
Fig. 2.8	After Platform on Disaster Displacement (2022). Tukuraki Village Relocation Project Information Brief
Fig. 2.9	After Platform on Disaster Displacement (2022). Tukuraki Village Relocation Project Information Brief
Fig. 3.8	After UNOCHA (2020). The relocation of Rohingya refugees into the Bhasan char
Fig. 3.9	After UNOCHA (2020). The relocation of Rohingya refugees into the Bhasan char
Fig. 4.1	After United Nations High Commissioner for Refugees (2007). Handbook for emergencies
Fig. 4.8	After Architects for Society on Architizer (https://architizer.com/projects/hex-house/)
Fig. 4.9	After Architects for Society on Architizer (https://architizer.com/projects/hex-house/)

Fig. 5.5	After REACH, An initiative of IMPACT Initiatives ACTED and UNOSAT
Fig. 5.6	After Global Shelter Cluster (2021). Shelter Projects 8th Edition
Fig. 6.8	After Balbek Buearu (2022). RE: UKRAINE
Fig. 6.9	After Balbek Buearu (2022). RE: UKRAINE
Fig. 7.1	Photo by Simon0523 on Free Photo Domain (Pixabay)
Fig. 7.5	Photo by Julie Ricard on Free Photo Domain (Unsplash)
Fig. 7.6	After Bris P, Bendito F (2019). Impact of Japanese post-disaster temporary housing areas' (THAs) design on mental and social health
Fig. 8.1	Photo by Ahmed Akacha on Free Photo Domain (Pexels)
Fig. 8.5	After Archdaily. Liina Transitional Shelter
Fig. 8.6	After Archdaily. Liina Transitional Shelter
Fig. 9.1	Photo by Rachel Claire on Free Photo Domain (Pexels)
Fig. 9.2	Photo from Free Photo Domain (Vecteezy.com)
Fig. 9.3	Photo by Alex Tinca on Free Photo Domain (Pexels)
Fig. 9.4	Photo by Ahmed Akacha on Free Photo Domain (Pexels)
Fig. 9.5	Photo by Quang Nguyen Vinh on Free Photo Domain (Pexels)
Fig. 9.6	Photo from Free Photo Domain (Vecteezy.com)
Fig. 9.7	After Shigeru Ban Architects
Fig. 9.8	After Shigeru Ban Architects
Fig. 10.2	After Fratzscher and Junker (2015), Integrating refugees: A long-term, worthwhile investment, DIW Economic Bulletin
Fig. 10.3	Photo by Pok Rie on Free Photo Domain (Pexels)

List of Figures

Fig. 1.1 Global displaced population by year (2010–2019) 2
Fig. 1.2 Children in refugee camp in Syria 3
Fig. 1.3 Percentage showing major causes of displacement 5
Fig. 1.4 Returned refugee population by decade (1980–2019) 6
Fig. 1.5 Unhealthy living conditions in shelters for the refugees in Idlib, Syria 8
Fig. 1.6 Elderly woman in shabby refugee shelter 9
Fig. 1.7 Five pillars of sustainability 14
Fig. 1.8 Keeping refugee and migrant children learning, healthy, and safe is part of social sustainability 15
Fig. 1.9 Principles for sustainable design 16
Fig. 1.10 Projected internal climate migrants by 2050 18
Fig. 2.1 Maximum recommended distance from community facilities 23
Fig. 2.2 Various climatic conditions can determine the location of dwellings on or near a hill 24
Fig. 2.3 Shelter position in the sloped topography in different climatic zones. 25
Fig. 2.4 Slopes exceeding 10% will restrict the design of the emergency housing 27
Fig. 2.5 Existing flora and fauna should be preserved for views and microclimate 29
Fig. 2.6 Orientation of the shelters in the housing should follow the sun path 31
Fig. 2.7 Tree plantation strategies for the indoor thermal environment 32
Fig. 2.8 Axonometric view of Tukuraki Village 35
Fig. 2.9 Plan of Tukuraki Village 36
Fig. 3.1 Site plan showing shelters arranged in varying rows along the contour lines. 42
Fig. 3.2 Modular camp master planning 44
Fig. 3.3 Cluster camp master planning 46
Fig. 3.4 Mixed land use with amenities for long-term needs 47

Fig. 3.5 Street categories . . . 49
Fig. 3.6 Different species of tree plantation for different benefits . . . 50
Fig. 3.7 The connection between open space and semi-public facilities . . . 52
Fig. 3.8 Masterplan of Rohingya refugee transitional housing . . . 54
Fig. 3.9 Cluster plan of Rohingya refugee transitional housing . . . 54
Fig. 4.1 Contingency planning process . . . 63
Fig. 4.2 Steps showing the shelter design process . . . 68
Fig. 4.3 Shelter design generation using generative design tool (Grasshopper) . . . 68
Fig. 4.4 Generating camp settlement size using the generative design tool . . . 69
Fig. 4.5 Dividing the camp settlement into smaller modules using the generative design tool . . . 70
Fig. 4.6 Generating central amenities using the generative design tool . . . 70
Fig. 4.7 Final refugee camp design generated from computational design tool . . . 71
Fig. 4.8 Floor plan of Hex shelter . . . 73
Fig. 4.9 Generative Hex shelter cluster . . . 73
Fig. 5.1 Conventional and curved laying method for manholes . . . 79
Fig. 5.2 Pit latrine section . . . 81
Fig. 5.3 Schools should be positioned in a convenient location inside the camp . . . 87
Fig. 5.4 Marketplaces should be located in a secure location accessible to everyone living in the community . . . 89
Fig. 5.5 Infrastructure master plan for Rohingya refugee camp in Bangladesh . . . 91
Fig. 5.6 Shelter design (elevation and section) . . . 92
Fig. 6.1 Example of emergency shelter . . . 99
Fig. 6.2 Example of transitional shelter . . . 100
Fig. 6.3 Example of durable shelter . . . 101
Fig. 6.4 Example of semi-detached shelter . . . 102
Fig. 6.5 Example of detached shelter . . . 103
Fig. 6.6 Example of row shelter . . . 103
Fig. 6.7 Row shelter cluster . . . 104
Fig. 6.8 Conceptual plan of refugee shelter prototypes . . . 111
Fig. 6.9 Perspective of refugee shelter prototypes . . . 111
Fig. 7.1 Prefabricated refugee accommodation or asylum seekers hostel in Germany . . . 116
Fig. 7.2 Steps leading to the construction of a townhouse using panelized and modular prefabrication systems . . . 118
Fig. 7.3 Automated manufacturing process of prefabricated panels . . . 120
Fig. 7.4 Example of Plug and Play washroom unit . . . 122
Fig. 7.5 Prefabricated container shelter for refugees . . . 123
Fig. 7.6 Plan of prefabricated shelter . . . 125
Fig. 8.1 Young adult participating in self-help construction . . . 133
Fig. 8.2 Using modularity while designing shelters . . . 137

Fig. 8.3 Step-by-step diagram showing the assembly process 139
Fig. 8.4 Diagram showing basic construction technique for wood shelter. . . . 140
Fig. 8.5 Floor plan showing self-built Liina transitional shelter. 141
Fig. 8.6 Step-by-step assembly diagram for self-built Liina transitional shelter. 142
Fig. 9.1 Bamboo house 150
Fig. 9.2 Clay brick house 151
Fig. 9.3 Adobe house with thatched roof 152
Fig. 9.4 Emergency shelter tent built with plastic sheet 154
Fig. 9.5 Thatched roof house. 154
Fig. 9.6 Cement built house. 155
Fig. 9.7 Paper Log House by Shigeru Ban 161
Fig. 9.8 Axonometric diagram showing the construction detailss 162
Fig. 10.1 Steps showing the planning and decision-making process 169
Fig. 10.2 Possible scenario for economic growth by integrating refugees into a country. 173
Fig. 10.3 Floods and droughts are common phenomena due to climate change 175

List of Tables

Table 3.1 Recommended module size. Source: (United Nations High Commissioner for Refugees 2007) 45
Table 4.1 Stages of emergency preparedness and response 60
Table 4.2 Modular planning units 69
Table 5.1 Minimum requirement for sanitary facilities (United Nations High Commissioner for Refugees. 2015) 80

Chapter 1
The Global Need for Shelters

Keyword Refugee crisis · Displacement · Humanitarian architecture · Refugee camps · Climate refugees · Emergency shelters · Forced migration · Internally displaced persons (IDPs) · Sustainable emergency housing · Disaster displacement

In the twenty-first century, a predominant global issue occupying the world's attention is the ongoing enormous refugee crisis due to wars, environmental disasters, or persecution. According to Devictor and Do (2017), the average period of exile for forcibly displaced people at the end of 2015 was 10.5 years (Devictor and Do 2017), during which many refugees and Internally Displaced Persons (IDP) spend their entire lives in camps. As a result, camps are increasingly being seen as long-term settlements rather than temporary housing. Regrettably, this necessity is mainly neglected by the planners of humanitarian organizations during camp planning. Regarding shelters, which are considered a fundamental human necessity, refugees are often offered inadequate or no semi-permanent infrastructure and facilities. As a result, camp dwellers are often forced to live in crowded and unhealthy living conditions for years. This chapter illustrates the global need for emergency shelters. First, an overview of global displacement will be given. This is followed by exploring long-term issues imperative for this field today. In closing, a framework of sustainable design principles will be outlined to propose a way forward.

1.1 Global Overview of Displacement

Over the last 10 years, at least 100 million people have been forced to leave their homes, seeking shelter either within or outside their country's borders. In recent years, forced migration and statelessness have remained high on the international agenda, generating dramatic headlines in every corner of the globe. According to available statistics, as of the end of 2023, the number of forcibly displaced people worldwide reached 117.3 million (UNHCR 2024a). This includes refugees, internally displaced people, asylum seekers, and others in need of international

A. Friedman, B. Chaki, *Fundamentals of Planning and Designing Sustainable Post-Disaster Shelters*, https://doi.org/10.1007/978-3-031-83317-5_1

protection. The number has continued to rise in 2024, with estimates suggesting that it has surpassed 120 million by April 2024 (UNHCR 2024b) (Fig. 1.1).

As of 2024, the situation for unaccompanied and separated children (UASC) remains critical. The number of UASC applying for asylum has continued to rise, with over 30,000 new applications recorded in 20,231. Additionally, there are now approximately 170,000 unaccompanied and separated children among the global refugee population (UNHCR 2024b). These children are growing up in refugee camps without adequate living standards, health care, education, and recreational facilities (Fig. 1.2). Hence, the refugee camps need other facilities for human life to live in the host country.

Climate change is one of the most critical issues of our time, and disaster displacement is one of its most serious consequences. Whole populations are already experiencing the effects, but marginalized people in some of the world's most fragile and violence-prone countries are adversely impacted. Climate change and natural disasters will amplify risks that cause citizens to leave their homes or cross international borders. Climate change, war, hunger, poverty, and persecution all interact to generate increasingly complex emergencies. According to the World Bank, by 2050, three regions (Latin America, Sub-Saharan Africa, and Southeast Asia) will produce 143 million additional climate migrants (Rigaud et al. 2018). As of 2024, the number of people displaced due to climate-related disasters has reached approximately 30 million (UNHCR 2024b). This includes those who have been forced to leave their homes due to extreme weather events such as floods, storms, and droughts, which are becoming increasingly frequent and severe due to climate change. Although the remaining two-thirds of displacements are due to other humanitarian crises, it is becoming clear that climate change is playing a role in so-called slow-onset events like desertification, sea-level rise, ocean acidification, air

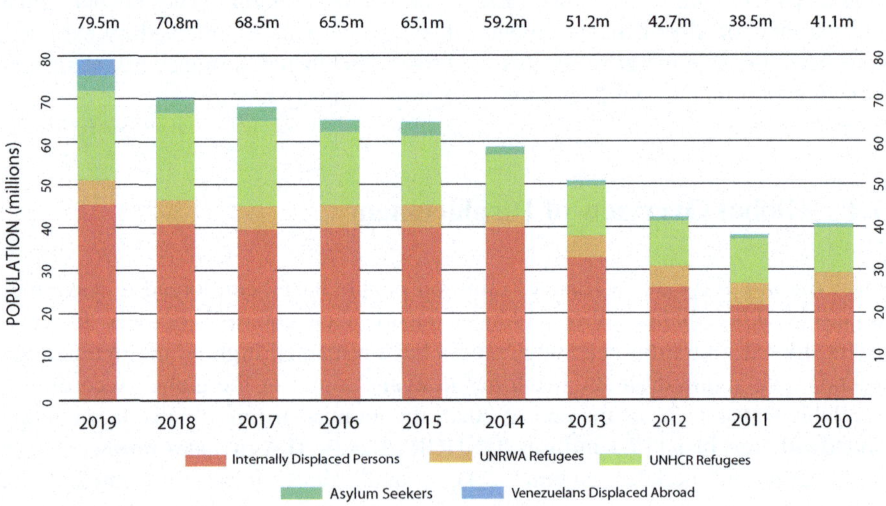

Fig. 1.1 Global displaced population by year (2010–2019)

Fig. 1.2 Children in refugee camp in Syria

pollution, rain pattern changes, and biodiversity loss. This deterioration will exacerbate many humanitarian problems, and more people will be forced to flee their homes. Following an environmental disaster, climate migrants require shelter as an immediate need. As the number of climate migrants grows, it will become increasingly difficult for humanitarian organizations to provide emergency shelter.

The cost of integrating refugees from other countries burdens many countries as the government needs extra money for the accommodation, social spending, and education of the refugees coming to that host country. In many countries, sub-central governments bear the burden of the cost of integrating refugees. Governments, the UN, shared financing channels, intergovernmental organizations, and the private sector contribute almost entirely to UNHCR for refugee settlement. UNHCR works all year to raise money for its programs and responds to new disasters when they arise (The UN Refugee Agency n.d.-a, -b).

1.2 Terminology

Many different terms used to describe refugees and immigrants, some with particular legal meanings. Properly using terms is essential for treating people with respect, understanding this global crisis, and searching for sustainable solutions.

- **Refugee**: A refugee is someone who has been forced to flee his or her country because of persecution, war, or violence (The UN Refugee Agency n.d.-a, -b).

- **Internally Displaced People (IDP)**: An internally displaced person is someone who has been forced to flee their home but never crossed an international border (The UN Refugee Agency n.d.-a, -b).
- **Shelter**: A habitable covered living space, providing a secure, healthy living environment with privacy and dignity to the groups, families, and individuals residing within it (Corsellis and Vitale 2005).
- **Transitional shelter/ resettlement**: Transitional shelter refers to a variety of shelter choices that aid individuals impacted by conflict or natural disasters who have been displaced or abandoned their homes, until they can find suitable permanent housing or return to it (Corsellis and Vitale 2005).
- **Refugee camp**: A refugee camp is a temporary settlement or sheltering area established to provide accommodation, basic services, and protection to individuals or groups forced to flee their homes due to conflict, persecution, natural disasters, or other life-threatening situations (The UN Refugee Agency n.d.-b).
- **Disaster displacement**: Disaster displacement pertains to instances in which individuals are compelled to abandon their homes or usual living areas due to a disaster or to pre-emptively escape the repercussions of an imminent and predictable environmental threat. It encompasses the adverse effects of climate shifts and disasters prompted by human actions like extensive industrial mishaps. It is important to note that displacement originating from conflict is not categorized under the scope of disaster displacement (Platform on Disaster Displacement n.d.).

1.3 Determinants of Displacement

The determinants of refugee situations are undoubtedly multifaceted and can vary substantially based on numerous political, economic, social, and environmental aspects. Here are several significant criteria that contribute to the existence of refugee situations:

- **Political instability and conflict**: One of the key drivers of refugee situations is political instability and conflict in the nation of origin. Wars, civil unrest, persecution, and human rights violations compel individuals and families to depart their homes in pursuit of peace and security elsewhere.
- **Economic hardship**: Economic reasons such as poverty, lack of employment possibilities, and economic inequality can also force people to become refugees. When people cannot pay for their basic requirements or enhance their level of living due to economic problems, they may be pushed to seek better possibilities in other nations.
- **Environmental factors**: Environmental calamities, including natural disasters such as floods, droughts, storms, and earthquakes, as well as human-induced environmental degradation such as deforestation and climate change, can displace communities and contribute to displacement. These environmental

variables often intensify existing social and political conflicts, leading to greater relocation.

- **Persecution and discrimination**: Refugees often escape their nations due to persecution based on criteria such as race, ethnicity, religion, political opinions, or membership in a particular social group. Discriminatory legislation, violence, and human rights abuses target specific people, pushing them to seek sanctuary elsewhere.
- **Lack of access to basic services**: In some situations, refugees may be driven from their homes due to a lack of access to fundamental services such as healthcare, education, and clean water. Conflict and economic instability can interrupt infrastructure and government services, leaving populations vulnerable and without crucial resources.
- **Social and cultural factors**: Social and cultural factors, including gender inequity, discrimination against marginalized groups, and restrictive social norms, can also contribute to refugee situations. In some situations, individuals may flee their country to avoid persecution or prejudice based on their gender, sexual orientation, or other parts of their identity.

These determinants interact and intersect in complicated ways, impacting the experiences and trajectories of migrants worldwide. Addressing the core causes of refugee situations demands a comprehensive approach that covers political, economic, social, and environmental aspects and emphasizes the protection and well-being of displaced populations (Fig. 1.3).

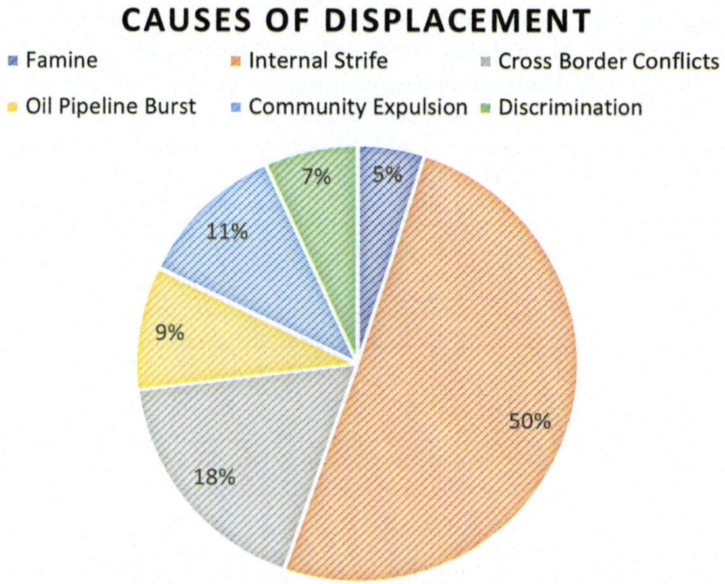

Fig. 1.3 Percentage showing major causes of displacement

1.4 Need for Long-Term Planning

Global refugee displacement stems from a complex interplay of multifaceted factors, reflecting the convergence of conflict, persecution, environmental challenges, and socioeconomic disparities. Armed conflicts and political instability in various regions force people to flee their homes in search of safety and security, resulting in one of the primary causes of displacement. Persecution based on ethnicity, religion, or political beliefs further contributes to this crisis, as individuals are compelled to escape life-threatening situations. Environmental factors, including natural disasters and climate change, have emerged as potent drivers of displacement, as communities are uprooted due to resource scarcity, rising sea levels, and extreme weather events. The multifaceted nature of global refugee displacement underscores the necessity of comprehensive and cooperative efforts to address these root causes and provide effective solutions for those in dire need of refuge.

Following displacement, many citizens are willing to return to their homes or pursue other options, such as voluntary repatriation or resettlement in third countries. Still, many more cannot do so and join the displaced population from previous decades. When wars and conflicts continue, fewer refugees and internally displaced people can return home. Other countries only allow a small number of refugees for resettlement, and host countries struggle to provide long-term needs for the displaced populations. Between 2010 and 2019, only 3.9 million refugees could return to their home countries. This compares to over 10 million refugees returning home in the previous decade and over 15 million two decades ago (Fig. 1.4). As more people are displaced and fewer can return, many people are forced to live in prolonged and long-term displacement circumstances. The world has transitioned from a decade of solutions to a decade of fresh and long-term displacement. According to Filippo Grandi, UN High Commissioner for Refugees, "We are witnessing a changed reality in that forced displacement nowadays is not only vastly more widespread but is no longer a short-term and temporary phenomenon" (UNHCR 2020).

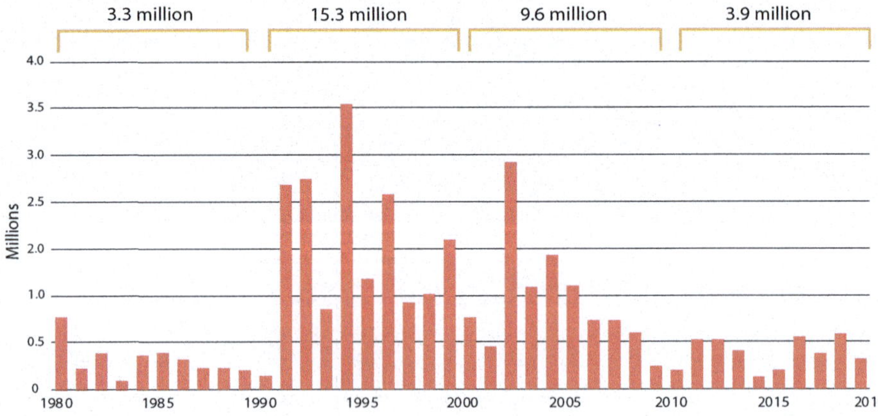

Fig. 1.4 Returned refugee population by decade (1980–2019)

Refugee camp settlements are quickly built under strict UNHCR guidelines to temporarily house displaced people due to conflicts. These grid layouts are adequate and practical for the short term, and they house people who have left their lives behind (Harrouk 2020). The grid camps resemble military camps in appearance and are often built by military engineers using "ordinary" military site-planning plans. Herein lies a significant issue: camps designed to meet military needs (i.e. a group of individuals) overlook essential criteria such as sufficient accommodation for families and the needs of particular refugee groups (Cuny 1977). In reality, for a large number of refugees, the camp has always been their only home for several years. Camps have struggled to address a permanent type of assistance tailored to the needs of refugees and host communities by basing their whole strategy on the premise that the situation is temporary and concentrating only on providing essential humanitarian requirements. To a degree, these spaces become unlivable as they transition from temporary to semi-permanent shelter. They rapidly become alienated slums as a result of numerous challenges such as a lack of adequate water supply, harsh environment, and poverty, among others. Since they are dense to minimize sprawl and appropriated land, these camps fail as they grow older and more extensive (Harrouk 2020).

One of the most basic human needs is a healthy built environment, which includes sufficient housing. Despite this, approximately one billion people live in slums and squatter camps, and a significant proportion of refugees live in insufficient shelters. Due to long-term stays in shelters made with fragile materials, the structure of the shelters and the quality of the environment inside the houses have become very poor and unusable (Habib et al. 2006). Several studies on refugee camps demonstrate that housing is often overcrowded and of poor quality. For example, a typical house in Sri Lankan refugee camps is a small wooden one-room hut with a corrugated iron roof. The dwellings are overcrowded, poorly ventilated, and lack a chimney (Turner et al. 2009). An analysis of houses in Jalazone, a Palestinian refugee camp, documented the unsafe housing conditions and discovered that 72.5% dampness, 50.5% had mould, 37% had leaks, and just 41.5% were exposed to light. Furthermore, 61% of households in Jalazone had 3 to 5 people living in each room, while in 16.5% of households, more than 5 people lived in each room (Al-Khatib and Tabakhna 2006). In the Rohingya refugee camp in Bangladesh, the density of the refugee camp is less than 10 sq. m. (100 sq. ft.) per person, which creates conditions comparable to the worst urban slums in Dhaka. Moreover, because of inadequate access to life-saving facilities in many areas, alerts from the health sector were becoming increasingly vociferous (Chaki and Friedman 2021). Malaria is a severe threat to both adults and children in refugee camps, especially where people live in close quarters with no window screens or solid doors (Unite for Sight n.d.). Various symptoms and illnesses have been linked dampness and mould, including aches and pains, digestive problems, and respiratory tract infections (Turner et al. 2009). Acute respiratory infections and poor mental health in children have also been linked to crowded, cramped conditions (Habib et al. 2006). A study conducted in Sri Lanka revealed inadequate living conditions in refugee camps, finding that living in a temporary camp was a significant risk factor for coughs, stomach pains, headaches, and

Fig. 1.5 Unhealthy living conditions in shelters for the refugees in Idlib, Syria

general malaise (Turner et al. 2009). Poor housing is also linked to a higher prevalence of rodents, which increases the risk of Lassa fever in West African countries (Bonner et al. 2007) (Figs. 1.5 and 1.6).

To avoid infectious diseases and epidemics while maintaining good health and dignity, adequate sanitation systems are essential and will be discussed in Chap. 5. Although the value of providing adequate latrines has been well established, most of the refugee camps lack adequate waste management or latrines. According to a study conducted in Bangladeshi refugee camps, camps with sanitation facilities had cholera rates of 1.6 cases per 1000 people. In comparison, camps without such facilities had cholera rates nearly three times higher (4.0–4.3 cases per 1000 people) (Cronin et al. 2008). According to a study conducted in a Kenyan refugee camp, sharing a latrine with three or more households is a significant risk factor for cholera because it increases the disease's faecal–oral transmission. In addition, the refugee camp must include facilities such as health centres, hospitals, schools, markets, distribution centres, and refuse drums for their long-term stay (Pascucci 2017).

Most camps are built quickly in response to mass relocation, and many are designed incrementally, even retrospectively, to bring order to a chaotic, improvised camp layout. Even though the initial strategic planning includes water and sanitation, housing, irrigation, access roads, and physical locations for providing health, education, and other services, it is difficult to consider the camp's entire lifetime during the design phase; political opposition, budget constraints, and a lack of consensus on duration are key factors that discourage longer-term planning. In many cases, displaced people are confined to the poorest parts of the country, far from host communities, cut off from facilities, and with few natural resources. This reduces

Fig. 1.6 Elderly woman in shabby refugee shelter

the likelihood of convergence with current programs and puts a longer-term financial burden on donors to finance settlements that cannot recover without ongoing support. Coordination of long-term investments from humanitarian and development actors may allow strategic planning of a camp, emulating traditional urban planning approaches as much as possible if political agreements are possible and resources exist for the gradual development of a camp into a permanent settlement. Approaches to spatial planning that view a refugee settlement as a "node"—linked to the physical, social, and economic existence of neighbouring territories—rather than an "island"—are helpful in long-term planning and strategically organizing subsequent interventions. For any long-term solution, the interests of the refugees and host communities must be considered. Early in the development plan, joint planning mechanisms must be formed between refugees, host communities, and governments to ensure that social stability and resource management are at the forefront of planning decisions. To ensure a successful process and outcome, effective consultation and interaction that connects policy, process, and spatial planning with the long-term needs of the settlement will need to be integrated (Moore 2017).

1.5 Social Aspects

Refugees face numerous socioeconomic challenges upon arriving in a new place, which can significantly impact their ability to integrate and thrive. One of the primary challenges is economic instability. Many refugees arrive with limited financial resources and face significant barriers to employment, such as language barriers, lack of recognition of their qualifications, and unfamiliarity with the local job market. This often results in underemployment or unemployment, forcing refugees into low-paying, precarious jobs that do not match their skills or experience (Byrd 2022). Additionally, refugees frequently encounter housing difficulties. Affordable housing is often scarce, and refugees may face discrimination from landlords, leading to overcrowded or substandard living conditions (Byrd 2022).

Social integration is another significant hurdle. Language barriers can impede communication and access to essential services such as healthcare, education, and legal assistance. Without adequate language skills, refugees may struggle to navigate daily life and build social networks, leading to isolation and mental health issues (Byrd 2022). Furthermore, cultural differences can create misunderstandings and prejudice, making it difficult for refugees to feel accepted and integrated into their new communities (Interiano-Shiverdecker et al. 2022).

Access to education is also a critical issue, particularly for refugee children. Interruptions in schooling due to displacement can result in significant educational gaps. In their new environment, children may face challenges such as language barriers, bullying, and a lack of tailored educational support, which can hinder their academic progress and future opportunities (Byrd 2022).

Healthcare access is another pressing concern. Refugees often arrive with physical and mental health needs stemming from their experiences of conflict and displacement. However, they may face barriers to accessing healthcare services, including language difficulties, lack of knowledge about available services, and financial constraints (Byrd 2022).

Overall, the socioeconomic challenges refugees face in a new place are multifaceted and interlinked, requiring comprehensive and coordinated efforts from governments, non-governmental organizations, and local communities to address. Providing targeted support in areas such as employment, housing, education, and healthcare is essential to help refugees rebuild their lives and contribute to their new communities (Graham and Miller 2021).

Technology significantly enhances the living conditions of refugees by addressing both their immediate needs and long-term social integration. Digital platforms facilitate access to essential services such as healthcare, education, and financial resources, which are often scarce in refugee camps. Mobile technology, in particular, enables refugees to stay connected with family and friends, preserving their social networks and providing emotional support (Aamira 2023). Additionally, technological advancements in housing solutions, such as modular and 3D-printed homes, offer more sustainable and quickly deployable shelter options (Aamira 2023). These innovations not only improve living standards but also foster a sense of community and stability among displaced populations. Furthermore, technology supports educational and employment opportunities, enabling refugees to rebuild their lives and contribute to their host societies (Culbertson et al. 2019). By integrating technology with social initiatives, we can create more inclusive environments that empower refugees to overcome their obstacles and achieve a better quality of life (Frey and Gatzweiler 2018).

Environmental conditions significantly impact the daily lives of refugees, often exacerbating their already precarious situations. Refugee camps are frequently located in areas with harsh climates, limited natural resources, and inadequate infrastructure, leading to challenges such as water scarcity, poor sanitation, and exposure to extreme weather (Ziersch et al. 2024). These conditions can result in health issues, including waterborne diseases and respiratory problems, further straining the limited healthcare resources available (Duclos et al. 2024). Social dynamics within refugee communities are also affected, as environmental stressors can lead to conflicts over resources and hinder community cohesion. However, community responses to these challenges often demonstrate resilience and adaptability. For instance, refugees have developed innovative solutions such as rainwater harvesting systems and community gardens to improve their living conditions. Sustainable solutions, such as the use of renewable energy sources and eco-friendly building materials, can further enhance the resilience of refugee housing. Integrating environmental sustainability with social initiatives can create more stable and healthy living environments for refugees, fostering a sense of community and improving overall well-being.

1.6 Psychological Aspects Effecting Refugees

The psychological impact of being a refugee is significant and complex, primarily arising from the trauma and stress linked to the displacement from one's home. Refugees frequently flee circumstances characterized by severe violence, warfare, persecution, or natural calamities, all of which can result in enduring emotional trauma (World Health Organization 2021). A multitude of refugees endure Post-Traumatic Stress Disorder (PTSD) due to their recurrent exposure to violence, bereavement, or the devastation of their residences (World Health Organization 2021). PTSD presents with symptoms like flashbacks, nightmares, and intense anxiety, which may be elicited by recollections of the traumatic incidents endured. Complex trauma, stemming from repeated and prolonged exposure to traumatic situations, frequently hinders rehabilitation, complicating the healing process for refugees.

Social pressures may increase the possibility of getting mental health disorders. Unemployment, adverse socioeconomic situations, and insufficient social integration among migrants and refugees are risk factors for mental health disorders, such as depression and anxiety (World Health Organization 2021). Depression is a major psychological consequence experienced by refugees. The profound sense of loss, encompassing the loss of home, community, and livelihood, can engender intense sentiments of pessimism (World Health Organization 2021). Numerous refugees endure a deep sense of bereavement, frequently exacerbated by survivor's guilt—an emotion characterized by remorse for having survived when others did not. The extended uncertainty of being in temporary accommodations, such as refugee camps, along with the protracted asylum process, intensifies these sentiments of hopelessness. Refugees facing challenges in acclimating to unfamiliar situations, where they may lack proficiency in the language and comprehension of the culture, are prone to experiencing frustration, isolation, and inadequacy, which can exacerbate melancholy.

Chronic anxiety is another widespread issue among refugee populations. Many live in constant fear of what the future holds, as they navigate unfamiliar surroundings, legal uncertainties, and the possibility of being sent back to unsafe conditions. This anxiety is often heightened by hypervigilance, where refugees remain on high alert for potential threats even in safe environments, a symptom of PTSD. This hyperawareness can be mentally and physically exhausting, leading to feelings of chronic stress, which only worsens over time (World Health Organization 2021).

Kids are particularly susceptible to the psychological ramifications of being in refuge. Early-life trauma exposure can result in emotional, behavioural, and developmental issues such anxiety, anger, or trouble focusing (World Health Organization 2021). Children's mental health can be further harmed by the disruption of their education and consistent routines during displacement, which can leave long-lasting emotional scars that affect their general development (World Health Organization 2021). Furthermore, it is critical that children have access to supportive surroundings that aid in their recovery because they frequently find it difficult to comprehend and absorb the trauma they have encountered.

A major contributing factor to the psychological suffering that migrants endure is social isolation. Families and communities are frequently split apart as a result of displacement, breaking vital support systems. Despite the packed surroundings, life in temporary shelters or refugee camps may be extremely isolating due to the stress of displacement, language obstacles, and cultural differences that make it difficult to make new friends. Acculturation stress, or the emotional strain of assimilating into a new culture while attempting to maintain one's own customs and identity, is a common problem experienced by refugees (World Health Organization 2021). This strain can lead to a profound feeling of estrangement, which exacerbates anxiety and despair.

The enduring psychological repercussions of refugee status can be profound, particularly if not treated promptly. A multitude of refugees endure chronic stress, existing in a state of uncertainty for years regarding the possibility of returning to their homeland or establishing permanent residence elsewhere. The extended period of uncertainty, along with insufficient mental health services in numerous immigrant environments, may result in mental exhaustion and exacerbate emotional turmoil. Unresolved trauma can impact not just the individuals who endured it but also their offspring, a phenomenon referred to as intergenerational trauma (World Health Organization 2021). In such instances, the emotional burden of displacement is transmitted throughout generations, with children inheriting their parents' stress and coping strategies.

1.7 Planning for Sustainability: A New Approach

The growing necessity for sustainable emergency housing has become a global need. The approach towards refugee housing needs profound transformation. These transformations require a rethinking of old planning policies and strategies. We must introduce innovative designs to refugee communities, shelters, and camps. We will examine three domains wherein fundamental changes have taken place and affected planning systems:

1. The social domain of displaced people, including population ageing, the increasing number of newborn babies in the total number of refugees, and the increasing need for providing education to the children.
2. The cultural domain, which promotes local culture and integration between cultural varieties, as well as the environmental domain, which includes the negative effect of dense refugee camps and excessive cutting down of trees.
3. The economic domain covers long-term job scarcity and the inability to earn money.

To have these social, cultural/environmental, and economic domains considered in refugee communities that developed gradually and generally spontaneously from an emergency camp into a new town begs the question: How should the needed changes be ushered in? How should refugee camps be planned anew or retooled? As a philosophy and organizing principle, sustainability has been put forward as an answer to these questions.

The origin of the term "sustainable development" can be traced back several decades. In 1972, the United Nations Conference on the Human Environment in Stockholm addressed concerns about the earth's carrying capacity. The meeting served as a forum for the first international discussion on the relationship between ongoing environmental damage and the future of humanity. It was recognized then that population growth in some nations and overconsumption in others were causing ecological issues such as land degradation, deforestation, air pollution, and water scarcity. Years later, this reflection led to establishing an international initiative, the World Commission on Environment and Development (WCED). In their 1987 report, *Our Common Future*, sustainable development was defined as a process that "meets the needs of the present without compromising the ability of future generations to meet their own needs" (World Commission on Environment and Development (WCED) 1987). This definition established a conceptual approach to sustainable development, whereby any action must be pursued with its future effects in mind. The WCED also created a paradigm for development whose main anchors are as follows:

- A need for social equity
- Incorporation of fair distribution of resources within and among nations
- The need to resolve conflicts caused by development pressures and the environment

These priorities have become a standard for sustainable development initiatives and should be considered during new development projects.

The three original pillars of sustainable development emerging from this document were *social, economic,* and *environmental* sustainability; however, it soon became abundantly clear that *culture* and *governance* must be part of any sustainability initiative if it is to succeed. A sustainable approach to culture and governance dictates that practices must reflect and respond to the values and traditions of the sociopolitical context in which they are implemented. While international efforts in the past have not always considered these factors, they have proven essential to their success (or failure).

These five pillars of sustainable development (Fig. 1.7) provide essential guidance for planning and designing sustainable emergency housing—so let us consider each in turn, beginning with social sustainability. Social needs imply a broad, all-compassing concept that can be explained and interpreted in many ways. It might be best understood by way of example. For instance, when creating a sustainable education system for refugee children is an objective, ensuring that sufficient funds will be continually available is essential. Providing enough schools and educational facilities can contribute to educating the children growing up in the camp. It is, therefore, in the best interest of humanitarian organizations that the camps are designed with schools and playgrounds for children and young teenagers.

Cultural sustainability is equally vital. Promoting vernacular culture, preserving the local traditions of refugees and their communities, and facilitating integration between them contribute to society in direct and indirect ways. Most obviously, integrating local tradition with the refugee camp design can create harmony between host and refugee communities, allowing them to know each other's culture and, more indirectly, helping the refugees to minimize mental trauma from their past displacement from the location of their culture (Fig. 1.8).

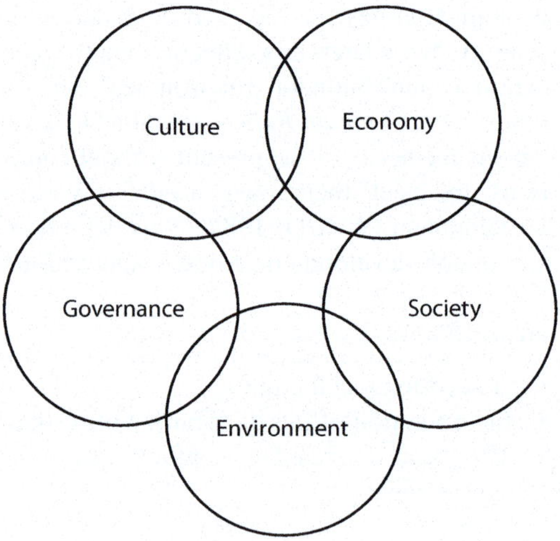

Fig. 1.7 Five pillars of sustainability

Fig. 1.8 Keeping refugee and migrant children learning, healthy, and safe is part of social sustainability

We cannot afford to ignore economic sustainability. Lousy planning decisions today burden the generations of tomorrow. Building temporary tent-like shelters, for example, will have long-term economic effects. Such shelters must be rebuilt after months or years, as they will not be livable. This requires more funding from the donor agencies to build and transfer the refugees from immediate emergency shelters to transitional shelters and then resettle them into permanent shelters. Due to this massive need for extra funds, in most cases, the refugees are forced to live in unhealthy and unstable structures.

Environmental sustainability concerns ecological attributes created by the construction and upkeep of development, including its roads, open spaces, and shelters. A "cradle-to-cradle" cycle assessment is necessary when planning a development. "Cradle-to-Cradle" is a term used in life cycle analysis to describe a material or product that is recycled into a new product at the end of its life so that there is no waste. This is in regard not only to the initial effect of choice of materials, for example, but also their long-term performance and recyclability. Unnecessary tree-cutting for a refugee settlement in host countries should be minimized. During the contingency planning stage, the sites' environmental conditions should be prioritized for selection, thereby considering environmental sustainability in planning and construction.

Lastly, we must consider sustainable governance. Planning strategies and concepts, innovative as they may be, will not be implemented unless a local leader of the host community allows and establishes appropriate policies. Political leaders and humanitarian organizations are equally responsible for implementing visions of long-term planning for refugees. Further, an effective political system will execute sustainable planning strategies for reducing long-term negative impacts on the local climate and economy due to accommodating the refugees.

Current methods of refugee housing face challenges due to their many far-reaching negative implications for society, the economy, and the environment. When an attempt is made to diagnose the root cause of poor planning practices, ignorance of one of four pivotal issues—environment, economy, culture, and society—can arguably be found at fault. Mainstream refugee settlements are often regarded as a temporary product rather than as a process, where a range of aspects are being systematically explored and manipulated. The process, the key issues, and the relations among them are illustrated with four general principles (Fig. 1.9). When followed, these principles can guide the conception of sustainable refugee housing.

- **The Path of Least Negative Impact**
  The first principle is the path of least negative impact. It dictates that decision-makers must minimize any negative impact the refugee camps might have on social, cultural, economic, environmental, or governmental practices. At the outset of any planning process, impact assessments should ensure limited short-term and long-term harmful impacts and costly ramifications. Implementing this principle is to establish an economic base that will not hinder future initiatives. For instance, permeable paving can infiltrate, treat, and store rainwater where rain falls, bringing cost-effective solutions where flooding or icing is a problem. Therefore, permeable paving in the refugee camp should be used to contribute to the return of water to nature.
- **Self-Sustaining Systems**
  Establishing a self-sustaining system is the second principle of sustainability. A self-sustaining system should be resilient, which is the capacity of a system to function in the face of outside change and pressure. For instance, providing land for agriculture can help attain long-term food sustainability. Therefore, while designing refugee housing, self-sustaining agricultural land with shelter should be considered from the beginning of planning.

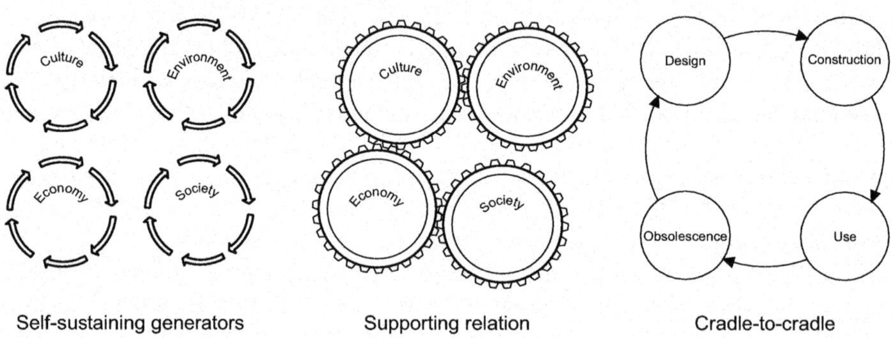

Fig. 1.9 Principles for sustainable design

- **Symbiotic Relationships**
  Another vital principle is the recognition of the symbiotic relationships between the five pillars noted above. Developing sustainable government, industrial, societal, cultural, and environmental programs will strengthen and support each other. For example, incorporating green open spaces in refugee housing will reduce climate change's impact, improving environmental sustainability. It will lower the cost of medical expenses by creating a healthy community, resulting in improved economic sustainability. Moreover, increased community participation and social interaction will boost social sustainability.
- **Life Cycle Approach**
  The final principle to consider is a life cycle approach. This means adopting a series of long-term policies and objectives. It advocates that while following their chosen path, a planner's overall strategy must remain flexible to accommodate ongoing changes as diverse needs arise. For example, choosing recyclable building materials for shelters can minimize landfill use, and in the long run, it can be beneficial for other purposes.

These five pillars critical to sustainable development can be viewed independently. However, when one closely examines the inner workings of refugee camps designed and built on sound sustainable principles, it is evident that the confluence of these aspects is critical. All integrate to affect the built environment; they operate in an elaborate balance. It is this complex nexus that this book intends to clarify for the reader. While we have organized and illustrated these sustainability concepts separately to single out each importance, it is no less than essential to recognize the interrelationships that connect them and see them as overlapping issues. Additional general principles are, therefore, articulated in the following section to aid the recognition of these relationships in the emergency housing context. The following chapters focus on the subjects vital to building sustainable refugee camps.

1.8 Climate Change and Global Migration: The Urgent Need for Action

It is vital to grasp the severe consequences of failing to address climate change successfully. If the world does not take decisive action to limit carbon emissions and global warming, the effects will be severe and far-reaching. One of the most significant repercussions will be the intensification of migration and displacement on an unprecedented scale. According to the World Bank's Groundswell report, if immediate action is not taken to reduce global greenhouse gas emissions, climate change could force up to 216 million people living in six different world regions—Sub-Saharan Africa, South Asia, Latin America, East Asia and the Pacific, North Africa, Eastern Europe, and Central Asia—to relocate within their own countries by the year 2050 (Clement et al. 2021) (Fig. 1.10).

As temperatures increase and extreme weather events become more frequent and severe, we may expect to witness a drastic shift in the distribution of populations

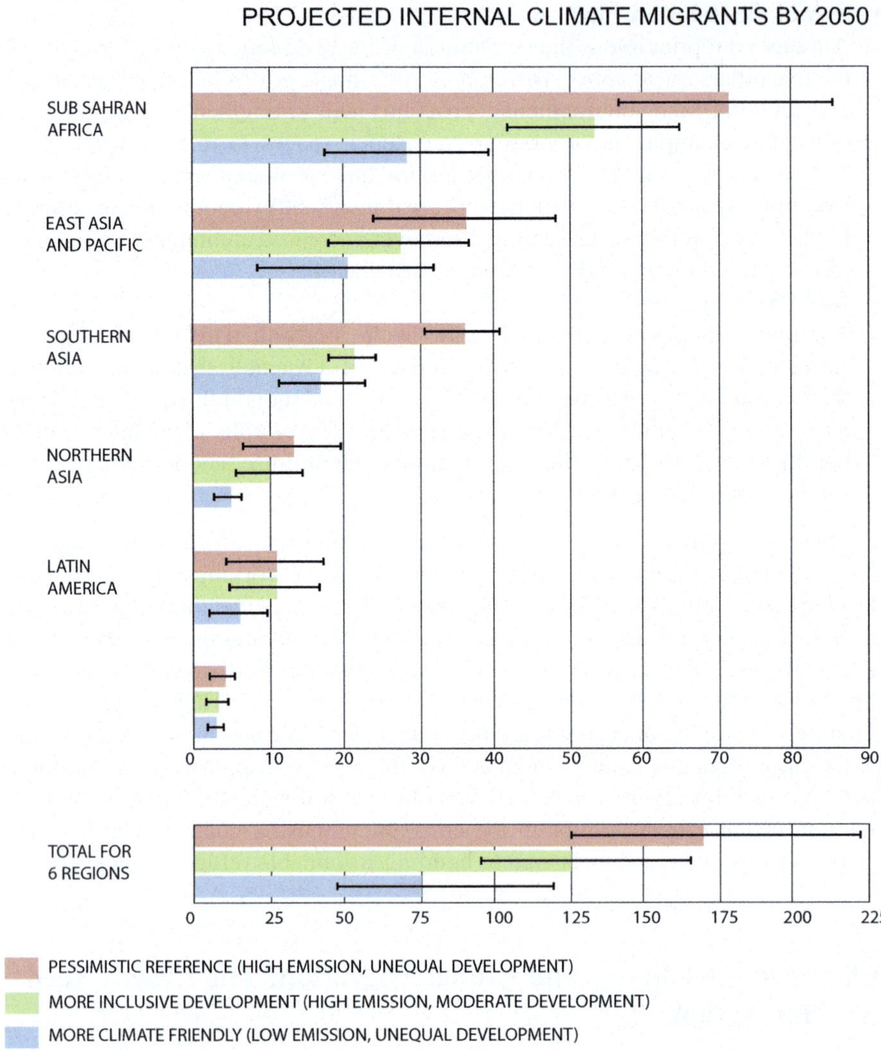

Fig. 1.10 Projected internal climate migrants by 2050

around the planet. Droughts, floods, and storms will not only cause damage to cities and infrastructure but also push individuals to flee their homes in search of safer places. What is particularly troubling is that these displacements will not be limited to the poorest and most vulnerable groups. As climate change undermines businesses and lives, even affluent towns and urban regions will be affected. The once-stable circumstances of affluent populations will no longer be immune to the consequences of environmental degradation.

The overwhelming number and complexity of these climate-induced migrations will need global solutions and coordinated efforts on an unprecedented scale.

Traditional techniques for managing refugee situations will no longer work in the face of such widespread displacement. Furthermore, the financial burden of reacting to climate-induced migrations would be enormous. The financial cost will be significant, from providing humanitarian help and refugee assistance to restoring infrastructure and adapting to new environmental conditions. Addressing climate change is not simply an environmental necessity but also a humanitarian and economic responsibility. Failure to take effective action now will only lead to greater pain, dislocation, and instability in the future. The global community must come together to develop practical solutions to reduce climate change and promote resilience in vulnerable populations.

Discussion Questions

- How do the current approaches to housing refugees fall short in addressing their needs?
- What are the potential long-term consequences of spontaneous construction of shelters without considering future needs?
- Why is there a constant need for emergency shelters around the world?
- What are some common approaches used by shelter organizations to address housing needs after a disaster?
- What are the long-term concerns associated with the design of emergency shelters, and how can they be mitigated?

References

Aamira F (2023) The use of technology to support the rights of refugees: an overview. In: IRIS Sustainable Development

Al-Khatib IA, Tabakhna H (2006) Housing conditions and health in Jalazone Refugee Camp in Palestine. East Mediterr Health J 12(1-2):144–152

Bonner PC, Schmidt W-P, Belmain SR, Oshin B, Baglole D, Borchert M (2007) Poor housing quality increases risk of rodent infestation and Lassa fever in refugee camps of Sierra Leone. Am Soc Trop Med Hyg 77(1):169–175

Byrd N (2022) Issues in refugee adaptation after resettlement. In: Why it matters. Utah Valley University, Orem, UT

Chaki B, Friedman A (2021) Housing the rohingya refugees in bangladesh: learning from the experiences of self-help emergency shelters cases. J Arch Des Construct Technol 2(1):1–10

Clement V, Rigaud KK, de Sherbinin A, Jones B, Adamo S, Schewe J, Sadiq N, Shabahat E (2021) Groundswell Part 2: acting on internal climate migration. Washington, DC

Corsellis T, Vitale A (2005) Transitional settlement: displaced populations. Oxfam-University of Cambridge :464

Cronin A, Shrestha D, Cornier N, Abdalla F, Ezard N, Aramburu C (2008) A review of water and sanitation provision in refugee camps in association with selected health and nutrition indicators—The need for integrated service provision. J Water Health 6:1–13. https://doi.org/10.2166/wh.2007.019

Culbertson S, Dimarogonas J, Costello K, Lanna S (2019) Applying technology to the global refugee crisis. In: RAND Social and Economic Well-Being

Cuny FC (1977) Refugee camps and camp planning: the state of the art. Disasters 1(2):125–143. https://doi.org/10.1111/j.1467-7717.1977.tb00020.x

Devictor X, Do QT (2017) How many years have refugees been in exile? Popul Dev Rev 43(2):355–369. https://doi.org/10.1111/padr.12061
Duclos D, Sharma E, Banke-Thomas A, Castaner MM (2024) Unsafe and unstable housing for refugees and asylum seekers is a threat to intergenerational health. BMJ 384. https://doi.org/10.1136/bmj.q561
Frey C, Gatzweiler M (2018) How tech can bring dignity to refugees in humanitarian crises. In: The conversation.
Graham J, Miller S (2021) How kenya can create shared growth by facilitating economic inclusion for refugees
Habib RR, Basma SH, Yeretzian JS (2006) Harboring illnesses: On the association between disease and living conditions in a Palestinian refugee camp in Lebanon. Int J Environ Health Res 16(2):99–111. https://doi.org/10.1080/09603120500538341
Harrouk C (2020) Refugee camps: from temporary settlements to permanent dwellings. In: ArchDaily. https://www.archdaily.com/940384/refugee-camps-from-temporary-settlements-to-permanent-dwellings. Accessed 14 May 2021
Interiano-Shiverdecker CG, Kondili E, Bazzi R (2022) Refugees in context: a systemic, intersectional exploration of discrimination in the United States. Int J Adv Counsell 44(2):263–282. https://doi.org/10.1007/s10447-021-09457-7
Mcdonnell T (2018) The refugees the world barely pays attention to. https://www.npr.org/sections/goatsandsoda/2018/06/20/621782275/the-refugees-that-the-world-barely-pays-attention-to. Accessed 30 Apr 2021
Moore B (2017) Refugee settlements and sustainable planning. Forced Migration Review
Pascucci E (2017) Community infrastructures: shelter, self-reliance and polymorphic borders in urban refugee governance. Territ Politic Gov 5(3):332–345. https://doi.org/10.1080/21622671.2017.1297252
Platform on Disaster Displacement (n.d.) Key definitions. In: Disaster displacement. https://disasterdisplacement.org/the-platform/key-definitions/. Accessed 18 Feb 2024
Rigaud KK, de Sherbinin A, Jones B, Bergmann J, Clement V, Ober K, Schewe J, Adamo S, McCusker B, Heuser S, Midgley A (2018) Groundswell: preparing for internal climate migration, Washington, DC
The Nansen Initiative (2015) Agenda for the protection of cross-border displaced persons in the context of disasters and climate Change, vol 1. https://disasterdisplacement.org/resource/nipa-vol1/. Accessed 1 June 2024
The UN Refugee Agency (n.d.-a) Government partners. https://www.unhcr.org/donors.html. Accessed 14 May 2021
The UN Refugee Agency (n.d.-b) What is a refugee? Definition and meaning. https://www.unrefugees.org/refugee-facts/what-is-a-refugee/. Accessed 30 Apr 2021
Turner A, Pathirana S, Daley A, Gill PS (2009) Sri Lankan tsunami refugees: a cross sectional study of the relationships between housing conditions and self-reported health. BMC Int Health Hum Rights 9(1):16. https://doi.org/10.1186/1472-698X-9-16
UNHCR (2020) Trends at a glance: global trends forced displacement in 2019. UNHCR The UN Refugee Agency: 1–84
UNHCR (2024a) Refugee data finder. https://www.unhcr.org/refugee-statistics/. Accessed 2 Oct 2024
UNHCR (2024b) Global Trends—Forced Displacement in 2023
Unite for Sight (n.d.) Food, water, sanitation, and housing in refugee camps. http://www.uniteforsight.org/refugee-health/module3#_ftn20. Accessed 30 Apr 2021
World Commission on Environment and Development (WCED) (1987) Our common future. Oxford University Press, Oxford
World Health Organization (2021) Mental health and forced displacement. World Health Organization. https://www.who.int/news-room/fact-sheets/detail/mental-health-and-forced-displacement
Ziersch A, Walsh M, Due C (2024) Housing and health for people from refugee and asylum-seeking backgrounds: findings from an Australian qualitative longitudinal study. BMC Public Health 24(1):1138. https://doi.org/10.1186/s12889-024-18616-5

Chapter 2
Site Selection

Keywords Site selection for refugee camp · Refugee camp planning · Displacement causes · Climate refugees · Forced displacement · Land use and refugee rights · Proximity to amenities for refugees · Site conditions for camp design · Accessibility and refugee health risks · Tukuraki village case study

The selection of a site for the planning and a refugee camp is primarily influenced by the cause of displacement—whether driven by climatic factors or forced migration. This determination is crucial in assessing the camp's permanence and the necessary services to support its residents. In most cases, for climate refugees, food and safe shelter are the priority to minimize loss of life following a disaster. Therefore, climate refugees are settled temporarily and later relocated to transitional shelters near the origin of living or other permanent locations. On the other hand, forcibly displaced refugees leave their country of origin and seek refuge in neighbouring countries. In such cases, refugees settled in a host nation are often sheltered in an undesired location. Although the site selection for forced refugees depends very much on the decision-making of the host country's government, their government officials or humanitarian organizations should consider a range of factors as the refugees tend to be in these camps for years.

This chapter delves into the critical factors that shape the choice of an appropriate location for various types of settlements, mainly focusing on the specific requirements for refugee camps and emergency housing. The chapter begins by emphasizing the significance of site size and area per capita, highlighting the correlation between overcrowding and health concerns. Land use and rights are explored, emphasizing the importance of legal agreements with local and national authorities for private land use. Proximity to amenities is discussed, underlining the advantages of locating camps near existing facilities to foster integration and prevent overburdening local resources. The impact of the site's natural condition, including slope, soil condition, groundwater, flood plains, flora, fauna, and microclimate, is comprehensively examined. The chapter also addresses orientation principles, considering sunlight and wind for passive design strategies. Additionally, it underscores the significance of local health risks, accessibility, and flexibility for future growth. A case

A. Friedman, B. Chaki, *Fundamentals of Planning and Designing Sustainable Post-Disaster Shelters*, https://doi.org/10.1007/978-3-031-83317-5_2

study of Tukuraki Village in Fiji demonstrates the practical application of these principles in achieving disaster-resilient housing and community development. The chapter offers valuable insights for effective site selection in diverse contexts through these considerations.

2.1 Size of the Site

A camp's size and area per capita are essential factors in site selection for planning because overcrowding contributes to increased morbidity and stress. Therefore, an essential requirement is sufficient space outside and within shelters. According to UNHCR, the recommended camp area per person is 45 sq. m. (484 sq. ft.), and the acceptable area per person is 35 sq. m (377 sq. ft.). According to the standards, a minimum of 45 sq. m. (484 sq. ft) per person should be allocated, including household gardening space when soil permits. All the infrastructure, such as roads, footpaths, educational facilities, sanitation, protection, firebreaks, administration, water storage, distribution points, markets, relief item storage, and plots for shelter, would all require 30 sq. m. (323 sq. ft.) per person. The remaining 15 sq. m. (161 sq. ft.) per person are allocated for household gardens on the family plot, which should be included in the site plan from the beginning. According to UNHCR standards, the unacceptable range is an area between 30 and 34 sq. m. (323–366 sq. ft.) per person, and the critical range is an area below 30 sq. m. (323 sq. ft.) per person (United Nations High Commissioner for Refugees 2015). While minimal area criteria for refugee sites are advised, they should be implemented cautiously and flexibly.

2.2 Land Use and Land Rights

Land for the establishment of refugee camps is scarce in the majority of countries. Sometimes, the government provides locations on public land. Any private land use must be based on formal legal agreements with the government and adhere to applicable national laws. Through an agreement with national and local (including traditional) authorities, the refugees should have limited use of the place. Traditional or customary land use rights are extremely delicate topics, and even if the national government agrees to utilize a place, local groups may object to its usage, even temporarily. Clarification of access rights and constraints on land use is also required to define the refugees' rights to gather fuelwood and timber, feed for their animals, graze their animals, and engage in agriculture or other survival pursuits.

2.3 Proximity to Amenities

Amenities is a catch-all term encompassing the services and the community support system a dweller needs daily and yearly. It includes daycares, schools, medical clinics, shopping, and services like a post office, a pharmacy, or a bank. Refugees are often provided with a neglected or remote site in their own region or the host country. However, for their long-term needs, locating their camp near existing amenities is highly viable. Moreover, the local population should not see the refugees as a burden because of their effect on existing local services and the environment. Instead, they should be provided with basic human facilities. Chapter 5 describes in detail the community facilities necessary for the long-term stay of the refugees in the camp. From the earliest stages of a crisis, suitable, well-selected sites and well-planned refugee communities with integrated, appropriate proximity to infrastructure are life-saving and alleviate misery. This might also prevent problems from being intensified. Furthermore, it may be more efficient in the long term, as there will not be additional problems regarding infrastructure to resolve. As a matter of principle, selecting a site with or adjacent to existing amenities will be desired. Designating an area or a place for such facilities for present or future needs would be wise. When a master plan for a refugee camp is prepared, proper distribution of such facilities can promote easy access for a significant number of users.

Figure 2.1 illustrates an essential aspect of site planning and design for communities, particularly in refugee camps and emergency housing. It visually represents the optimal distance between community facilities and residential areas. It highlights that these essential amenities, such as schools, medical clinics, and markets, should be strategically located within a certain distance from the residential spaces to ensure easy access and convenience for the inhabitants (De Chiara and Koppelman

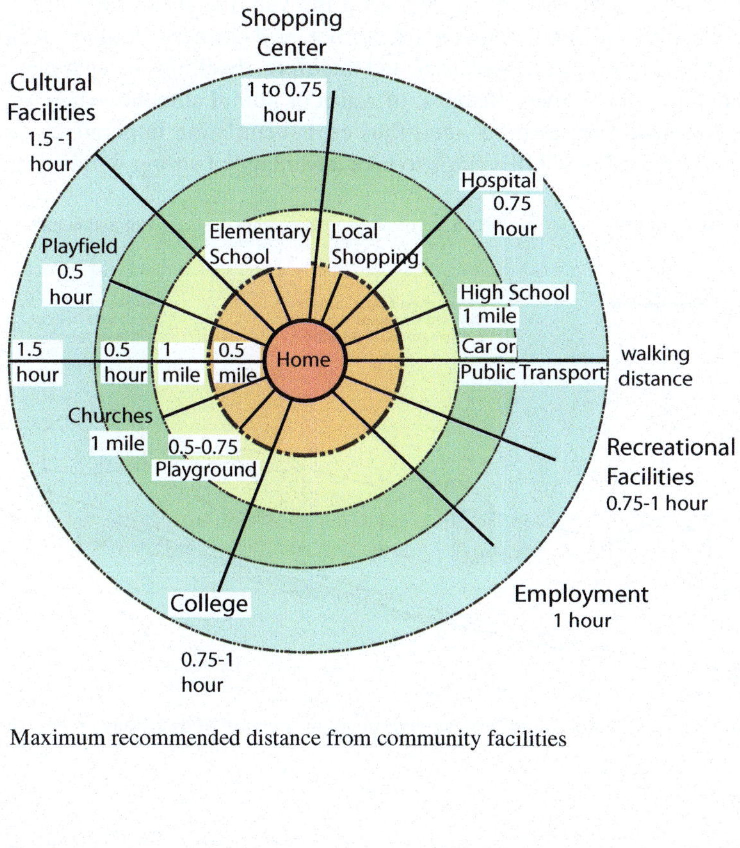

Fig. 2.1 Maximum recommended distance from community facilities

1975). This design approach aims to promote efficient use of resources and minimize the distances people need to travel to access vital services. The figure underscores the principle of integrated and well-planned communities, where proximity to amenities is pivotal in enhancing the quality of life and promoting social cohesion among residents.

Building on virgin land on the fringe of town can yield savings through a lower land cost, yet these savings may be offset when the cost of rudimentary infrastructure is factored in. This brings to the forefront the negative economic implications of sprawl. Once built, roads and other utilities will have to be maintained at mounting costs. Building denser neighbourhoods fosters environmental sustainability and helps keep costs down for the entire community by not having to service a larger area.

2.4 The Site's Natural Condition

When selecting a site for a refugee camp, the site's natural conditions need to be verified before a decision is made. It is crucial to consider the position of shelters in relation to any slopes or hills as this will largely determine the weather effects it will be subject to (Fig. 2.2) (Holden and Liversedge 2011). In cooler or temperate climates, it is essential to shelter the buildings from cold wind. Moreover, as wind velocity increases at the ridge of a hill and cold air sinks to the bottom of a valley, it is best to locate a house halfway between the valley and the ridge to avoid both undesired extremes. On the other hand, for hot, arid climates, cold air, which settles at the bottom of the valley, is desired, and therefore, dwellings in such places should be located at a lower level. Besides, in warm or humid climates where more wind and air circulation are desired—and thus cross-ventilation imperative—the dwelling should be sited on a hill's ridge to take advantage of strong winds (Fig. 2.3).

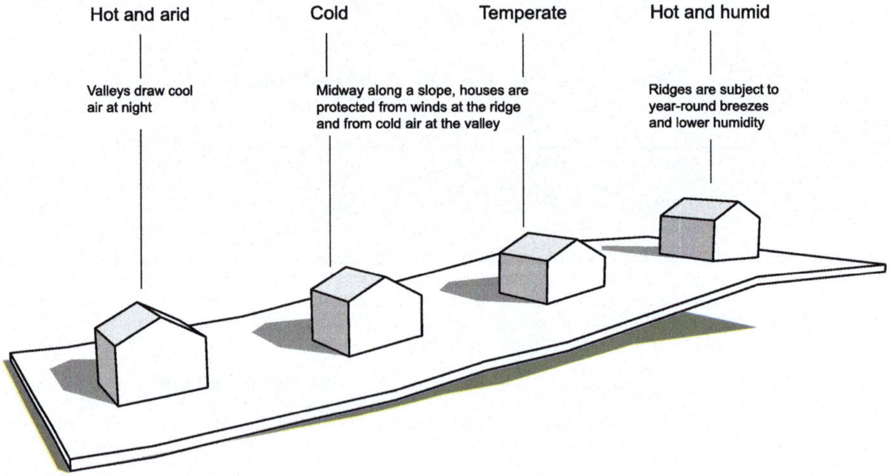

Fig. 2.2 Various climatic conditions can determine the location of dwellings on or near a hill

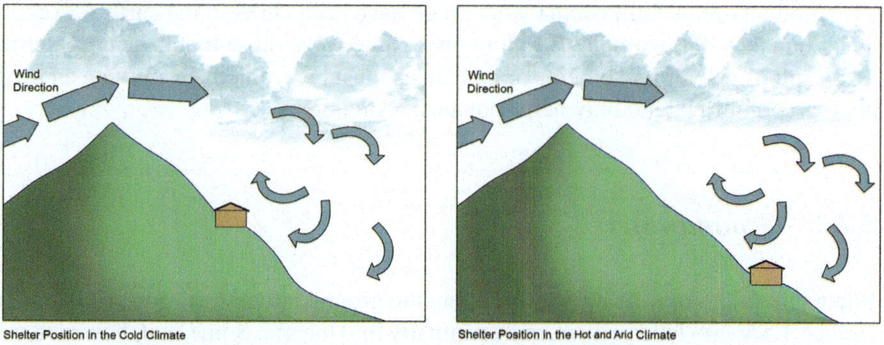

Fig. 2.3 Shelter position in the sloped topography in different climatic zones

Generally, a building should adhere to the site's existing natural conditions and fit accordingly. Finding a suitable site by considering all environmental factors relating to elevation and topography also means considering the effects and potential gains from weather conditions, such as wind, light, or temperature. Aspects such as soil composition, topography, presence of groundwater, vegetation, and orientation will affect design and cost. In an emergency housing project, these issues ought to be given greater priority. The following discussion is meant to help the planners and designers of refugee housing projects verify what they should be concerned with when selecting a site.

2.4.1 *Soil Condition*

The word soil is commonly used to describe all the fragmented materials in the earth's crust. It includes materials ranging from sand and gravel to fine clay derived from solid rock. Each region has its own type of soil, which requires engineers to adapt foundation designs and backfill methods to suit the characteristics of these soils (Duncan 1992). When considering a building site, an initiator should not be deceived by its initial appearance. Soil-related issues can still arise at a greenfield site that has not been disturbed. With increasing land costs, small building firms often purchase a leftover lot, not realizing that more expenses will be needed when design and construction begin.

In general, avoiding sites with soil that will not adequately support a building through its life or that may damage a structure through soil movement or flooding makes sense. Before deciding on a site, it is essential to follow some steps to ensure there is nothing wrong with the location. If the site remains vacant in mostly populated areas, suspicion needs to arise as to the reason. Professionals need to be consulted in rural areas, and soil testing may be conducted. Information can also be obtained by inquiring about the site's history.

When considering soil types suitable for construction, common knowledge is that rock does not usually present a problem in the construction of dwellings unless it rests on a fault. A fault occurs when there have been shifts in the earth's level due to earthquakes, for example. Building on organic soils, those that have been shifted as a result of glacier movements or landslides must be avoided. Examining an aerial photograph of raw land may help spot such a situation.

2.4.2 *Groundwater*

When rain falls, some of it will flow along the ground into a storm sewer, and on the surface, back into lakes, rivers, and eventually into the sea. Some of it, however, will filter into the ground. Other portions will be held close to the topsoil, yet most will flow until it reaches a larger body of water accumulating underground from previous rainfalls. This reservoir of water, which is held underground, is called groundwater.

The top of the underground body of water is also known as the water table. Whether in a contained state or flowing into a large body of water, having underground water on site will lead to costly consequences for foundation design and maintenance as it will affect the foundation's stability. Therefore, if the water table is elevated, the choice of the site must be reconsidered.

2.4.3 *Flood Plains*

Sites located in floodplain areas may be, at times, attractive to shelter organizations for refugee housing. However, the choice of building on them should be carefully considered before a decision to purchase the site is made and preferably strongly avoided. A flood zone is an area located adjacent to a river or lake. In springtime, when snow and ice melt rapidly or when there is heavy rainfall, the area may become flooded (Gilmer 2019). Flood-prone areas can be described as one-zone or two-zone. The one-zone is the land that can be covered by water in a once-in-a-hundred-year flood that can effectively happen at any time. The two-zone is where the water streams on its way to zone one.

Each year, extensive and costly damage is reported in the Kutupalong refugee camp in Bangladesh due to flooding (International Organization for Migration 2019). Building refugee housing in such zones is not advisable because often the residents cannot afford the necessary repairs, and they will be bound to lose everything. Most municipalities and national authorities have maps that designate areas with the greatest building risks and lands to be avoided altogether. Before selecting a site, it would be wise to consult such documents.

Fig. 2.4 Slopes exceeding 10% will restrict the design of the emergency housing

2.4.4 *Slopes*

At times, when the site is in a hilly area, there is no option but to build on slopes, which provide an opportunity for creative design through views and grading. There are times, however, when the slopes exceed a particular norm, leading to additional expenses in fitting the buildings to the site. Slopes of 0–8% are generally considered optimal for roads and pedestrian paths. About 2–8% grades work best for buildings if appropriate grading measures are taken for drainage. Slopes of 5–10% will be more suitable for small, detached homes due to the planning flexibility that they provide. When the slope exceeds 10%, the design will be more restricted (Fig. 2.4). When larger structures are considered for the site, fitting them will be more expensive. The homes must be "broken up" according to the elevation, leading to more complicated and costly designs. One can also expect higher development costs for infrastructure like roads, which are more complex to build and adequately drain in elevations (Wietek 2022). In landscaping plans, one must also ensure that rainfall from one lot will not flow into an adjacent lot. Therefore, it is recommended that when an affordable refugee project is contemplated for a site with a significant slope, the costs of fitting the design to the site should be figured out in advance.

2.4.5 *Brownfield Sites and Wetlands*

The term Brownfield site refers to lands previously used by industry. Activities on those properties might have contaminated the soil or the building, be it a vacant lot or an abandoned building. Soil testing and auditing of the site according to standards set by environmental agencies will most likely be needed before new uses can be proposed. Building on such a site will require replacing the topsoil to a depth determined by a test, which can be costly (De Sousa 2021).

In addition to the site's soil composition and grading, the natural coverage, also called wetlands, will affect development costs. Whether adjacent to salt or freshwater, wetlands are broadly defined by the vegetation found on them, not by the presence of standing water. In the past decade, wetlands have become all that remains for building in many areas. These sites hold the potential for attractive community design due to their natural beauty. By law, wildlife and trees are suggested to be kept and altering the terrain is usually prohibited. Preserving a portion of the land may increase the cost of each dwelling unit, making it unaffordable.

On the other hand, since shelter organizations are reluctant to develop wetlands sites, such sites may be accessible from host countries. Creative design can take advantage of the site's natural conditions. For example, higher-density housing may be built on a portion of the site, leaving the rest as green public space. The site, therefore, needs to be examined carefully before the design proceeds.

2.4.6 Flora and Fauna

The existing flora and fauna play a crucial role in maintaining the ecological balance of an area. Preserving them ensures the survival of native species, contributing to biodiversity and overall ecosystem health. The presence of greenery and natural beauty has a positive impact on mental health (Yerro 2011). Refugees often face traumatic experiences; preserving existing flora and fauna can provide a sense of tranquility and respite. Figure 2.5 shows how to place shelter by preserving existing greenery and natural environment. The site should not be near ecologically, environmentally sensitive, or protected regions. When chosen, a site will preferably have ground cover (e.g., grass, bushes, and trees). Vegetation provides shade and also reduces erosion and dust. Care should be taken during site preparation to cause as little harm to plants and topsoil as possible. Indiscriminate bulldozing or topsoil removal must be avoided at all costs when using heavy equipment. If wood is required for domestic cooking or shelter construction, the refugees should be advised not to obtain supplies from the camp's surrounding neighbourhood. Instead a more dispersed pattern of wood collecting should be developed in collaboration with local forestry authorities. A short survey of available vegetation and biomass should be conducted to prepare an ecological conservation plan.

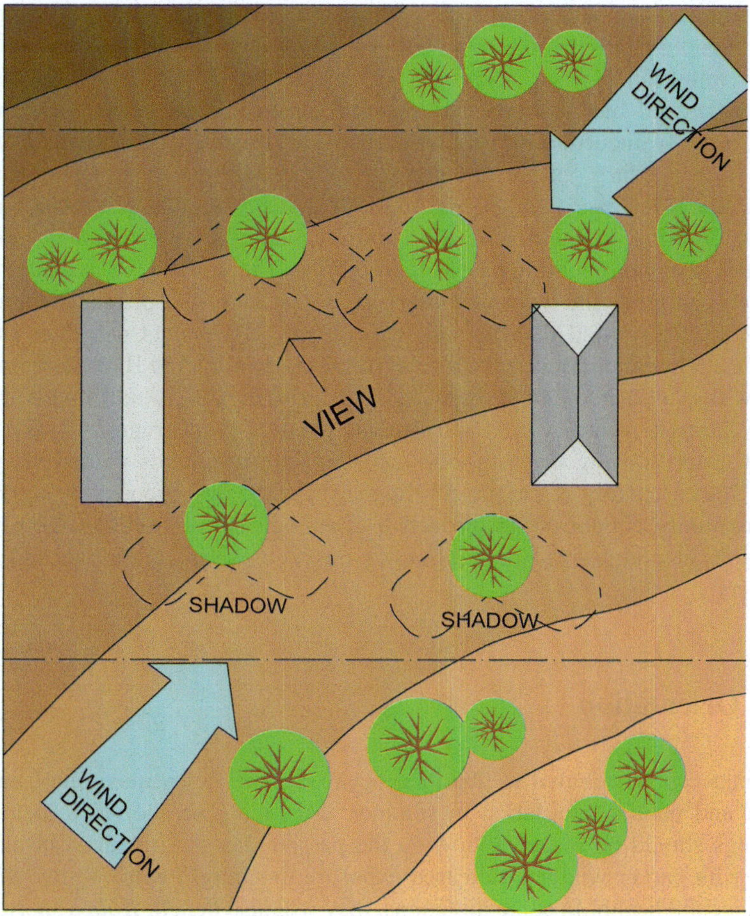

Fig. 2.5 Existing flora and fauna should be preserved for views and microclimate

2.4.7 Appropriate Microclimate

Each region has its unique climatic characteristics. The site may also be affected by factors associated with its location, orientation, and landscape. Appropriate planning may lead to energy conservation in each dwelling and, as a result, substantial savings in upkeep costs. Before selecting a site, the design team will have to figure out in which direction most shelters can be oriented. The East-West direction will expose the dwelling's longer elevation to the south sun, enabling passive solar gain (LaGro Jr. 2007). Wind direction can also be crucial, especially in sandy, arid areas, as the cold wind can lower the shelter's temperature in winter. When trees grow on the site, their importance in blocking drifting snow and wind can also be studied. Designs that will consider the site's microclimate can lead to money savings and greater comfort.

During the site selection process for post-disaster housing for internally displaced people (IDPs), conducting a thorough disaster risk assessment is of paramount importance. The aftermath of a disaster often leaves communities vulnerable, and hastily chosen relocation sites can exacerbate the challenges IDPs face. By incorporating a comprehensive disaster risk check into the site selection procedure, authorities can proactively mitigate potential hazards and ensure the safety and well-being of the displaced population. Assessing risks such as flooding, landslides, seismic activity, and other environmental threats is essential to avoid repeating past vulnerabilities. The knowledge gained from past disasters can guide the identification of safe areas less susceptible to further damage. A well-informed disaster risk check enables planners to select sites that provide shelter and reduce exposure to future risks, allowing IDPs to rebuild their lives with confidence and resilience. By prioritizing safety and long-term well-being through rigorous risk assessment, disaster-prone regions can establish post-disaster housing that supports recovery and safeguards vulnerable populations. The case study of Tukuraki Village (Sect. 2.9) underscores the significance of community collaboration, proactive planning, and comprehensive resilience measures in addressing the challenges posed by climate change and natural disasters.

2.5 Orientation

Buildings can be oriented so that their occupants will enjoy maximum sunlight in winter and natural ventilation in summer. Depending on the site's location, the buildings should be placed following the passive design strategies. In temperate zones with darker winters, increased exposure to sunlight will improve the occupant's mental health. In some directions, they will also benefit from a shorter noonday shadow than what would be cast if they were oriented east-west. Moreover, additional structures can be built upon the cusp of a neighbouring one's shadows if the design follows the layout mentioned above. Principles that consider sunlight and shadow orientation must avoid obstructions that will prevent buildings from maximizing sunlight (Atkinson 2018).

The location and orientation of buildings can maximize the capture of solar energy year-round, particularly during the winter season. A shelter should be oriented to receive the most direct sunlight during the day, with its elongated side facing south. It is imperative that the dwelling be oriented towards true south and not magnetic south, which can differ by up to 20° (Ahmed and Sulaiman 2003). To find true south, the location's declination must first be determined. The declination is the number of degrees east or west between compass (or magnetic) north and true north. Once the declination is decided on, the true south can be determined. Having the elongated side of the dwelling along the east-west axis means

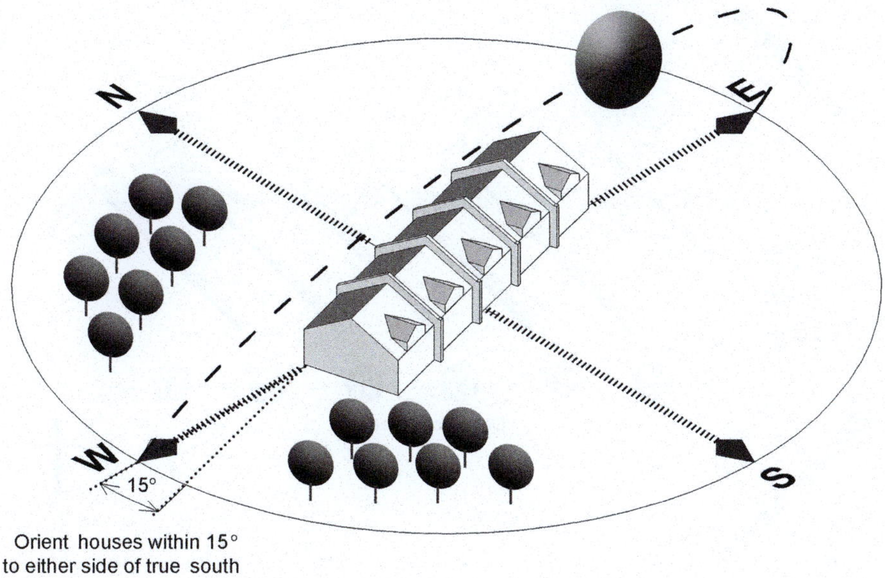

Fig. 2.6 Orientation of the shelters in the housing should follow the sun path

that it can be reached by the low-angle rays of the winter sun and conversely, placing the short sides of the building to the east and west to minimize solar gains during the hot periods of summer (Numbers 1995) (Fig. 2.6). Furthermore, the dwelling should be located in areas with unobstructed sunlight, thus, placing a dwelling too close to adjacent objects or buildings should be avoided. Not only should the elongated side be exposed to the south to benefit from this orientation fully, but more windows should also be placed along this axis, as the heat acquired through passive solar gain far exceeds any heat loss from the windows of that façade.

Incorporating the existing environmental elements of the site to promote natural ventilation is another essential aspect of sustainable design. Trees, vegetation, and landscape are crucial in determining wind flows. For example, near bodies of water, the wind will flow from the direction of the water towards the land in the daytime but will reverse at night. Because of inertia, the wind will flow around objects and keep the same direction of flow. As a result, the vegetation in the surrounding area can significantly reduce the wind speed. During the cold months, trees protect from the winter winds that generally originate from the north. Consequently, coniferous trees to the north of houses are appropriate as they maintain their foliage in the winter (Fig. 2.7). This also allows the shelters to benefit from south-eastern summer breezes. Considering the existing wind flows means that there will be less reliance on mechanical and resource-intensive heating and cooling systems, thus making the development more sustainable.

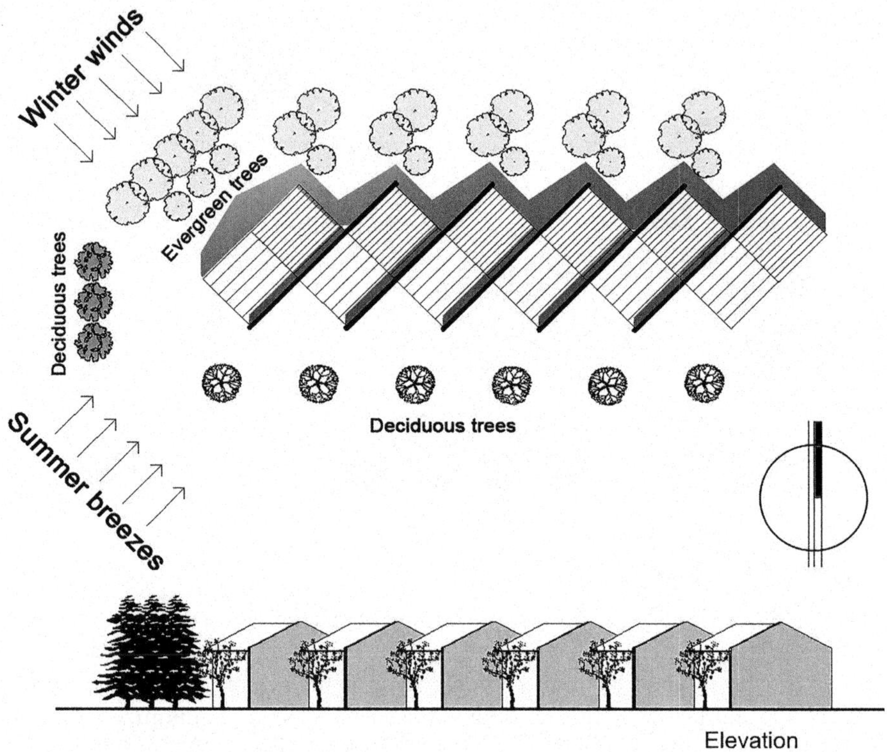

Fig. 2.7 Tree plantation strategies for the indoor thermal environment

2.6 Local Health and Other Risks

Settlement areas with major environmental health threats such as malaria, onchocerciasis (river blindness), schistosomiasis (bilharzia), and tsetse fly should be avoided for emergency housing. Unseen and irregular (though frequently localized) threats, such as flash flooding or severe industrial pollution, may exist at a site. Regular dust clouds can exacerbate respiratory illnesses at sites in dusty environments. While emergency and temporary shelters require protection from severe winds, a daily breeze is beneficial. Climate conditions should be adequate year-round, and seasonal variations should be considered. A suitable site during the dry season may become intolerable during the rains. Similarly, hilly places may be appropriate in the summer, but temperatures may dip well below freezing in the winter. Seasonal changes can significantly affect the type and cost of housing, infrastructure, heating fuel, and even cuisine. As much as possible, refugees should avoid being resettled in areas with significantly different climates than those they are accustomed to. For instance, resettling refugees from malaria-free high land in a swampy location infected with disease can be deadly (United Nations High Commissioner for Refugees 2007).

2.7 Accessibility

It is essential to consider the state of local roads and their closeness to transport hubs to distribute relief supplies and other items. Relief help must be delivered so that it does not wreak havoc on the local road infrastructure. The facility and any key storage and food delivery points must be accessible by heavy trucks through an all-weather road. Additionally, all other amenities must be accessible by light vehicles. Safe, secure roads and paths within settlements are required, as is all-weather access to all private residences and communal facilities. It is also crucial that humanitarian organizations consider the needs of individuals with mobility or access barriers (Sphere 2018).

The planners should ensure that the site has sufficient road infrastructure and safe and easy accessibility, even during the rainy season. Internal road networks, significant walkways linking the different usage areas, and local footpaths should all be included in each camp's road system (Cuny 1977). The road would provide refugees with direct access to trade and any informal information networks that may grow along the road. The camp's position along the road aids in dispelling the refugees' perception that they are being held isolated and apart from the rest of the world. Since the road is central but linear, the numerous larger non-residential buildings will be able to spread out, placing a greater proportion of the shelters closer to the "centre" while still maintaining a high level of accessibility (Cuny 1977).

2.8 Flexibility and Adaptability for Future Growth

During the emergency period after a conflict, the influx of people coming to the host region is hard to regulate. The size of the camp determined at the beginning of the site selection might not accommodate the incoming influx of refugees. When it comes to addressing the varied and ever-changing requirements of displaced populations, the social aspect of flexibility and adaptability in the expansion of refugee camps is absolutely necessary. As a result of the fact that refugees come from a variety of cultural backgrounds, family structures, and age groups, their physical and social requirements are also distinct. The feeling of being displaced can be made even more difficult by a camp structure that is tight and inflexible, which can increase tensions and discomfort already present. In refugee camps, the social dynamics are influenced by the degree to which the environment is able to adjust to the many demands of the people living there, and flexibility is an essential component in the process of cultivating a community that is both supportive and cohesive. As camps expand, flexibility allows for the provision of vital services, which contributes to the preservation of social order and normalcy. It is possible to assure both equity and protection by addressing the needs of vulnerable populations, such as women and children, through the utilization of flexible spaces. In addition, including refugees in the decision-making process about their living environment encourages community engagement, empowerment, and cooperation, so producing a social environment that is more democratic and helpful.

To fit the growth of refugee camps, flexibility is best defined in two ways: the ability to be easily modified and the willingness to change or compromise. The definitions denote essential aspects of flexibility that appear in a refugee camp. From a built environment perspective, flexibility is the possible movement and modification of the infrastructure and buildings, such as housing, water, sanitation, and hygiene (WASH) facilities, schools, and distribution centres. For instance, if WASH facilities are transformed from communal to private, this shows forms of modification and the camp's willingness to allow or accept this change. Flexibility as an architectural concept will be the primary tool for dominating these physical and social requirements within the constrained modular areas.

2.9 Case Study: Tukuraki Village, Fiji

Design Organization: The Building Safety and Resilience in the Pacific (BSRP).

Tukuraki, a Fijian village in the highlands of Ba, Viti Levu, had effectively relocated from a disaster-prone site to a new location, demonstrating the future of climate resilience, a reality faced by Pacific communities vulnerable to climate change and natural catastrophes. Their new homes were designed to withstand storms of Category 5. Tukuraki was struck by three natural disasters in 4 years. A landslide buried much of the village in January 2012, destroying a house where a family of four was sleeping, and damaged the settlement's roadways, water supply, and community hall (Building Safety and Resilience in the Pacific (BSRP) 2017).

Tropical Cyclone Evan (Cat. 4) devastated the village in December 2012, and Tropical Cyclone Winston damaged it again in February 2016. (Cat. 5). Before returning to their temporary homes, the people sought refuge in a cave and surrounding settlements (SPC Geoscience n.d.).

The suffering of the Tukuraki inhabitants forced the Fijian government to move the small community of around 40 people to a more disaster-resistant area. This was accomplished with the assistance of the Tukuraki community. After years of living in a temporary location, the community collaborated with the Ministry of iTauki Affairs, the Western Division Commissioner, the National Disaster Management Office, and other partners to obtain a permanent and safe site for the community, finalized in November 2015. The community spent 3 years securing a new plot of land that was both more resilient to catastrophe and suitable to the community for its safe migration. The community's previous site selection was examined by the Mineral Resources Department and determined to be in danger of future disaster, forcing the village Turaga in Koro (Chief) and the community to seek out and negotiate a new location (Building Safety and Resilience in the Pacific (BSRP) 2017).

The community was assisted by development partners, non-governmental organizations, and foreign agencies in obtaining the necessary permits and requests to access additional property inside their local district and securing a previously utilized agricultural site (Building Safety and Resilience in the Pacific (BSRP) 2017).

The new community's development has been directed by rigorous risk resilience, including GIS mapping of the new site development for the houses and community hall to Category 5 cyclone standards, rainwater collecting systems, and tanks that provide water and sanitation to the entire community. Additionally, sustainable livelihood initiatives have been developed, including a fishpond for food production, community economic empowerment, a poultry farm, 15 beehives for honey production, and women's economic empowerment projects such as handicrafts and weaving (Building Safety and Resilience in the Pacific (BSRP) 2017).

The new access roads to the village, drains, and culverts were constructed to prevent the risk of landslides. The construction of the dam, installation of 10 storage tanks, filters, and household distribution through piping for clean and safe drinking water were also completed as a part of the project. The mineral resources department advised on proper drainage, which was completed for the new village site. A 345-metre retaining wall prevents erosion on the steep edges of the community. Rainwater harvesting from rainwater collected from the roofs of new homes installed for all homes. A wastewater system was installed for all homes, leading to a soakage pit. Along with the eleven houses and evacuation centre, the Tukuraki village received assistance constructing road access, site levelling, and a playground. The new location is closer to Nalotawa District School, providing the community with improved access to education and health facilities.

The community and the Fijian government genuinely drove this community resettlement, and the outcome demonstrates their dedication. Tukuraki Village is now a symbol of disaster-resilient housing projects throughout the world (Figs. 2.8 and 2.9).

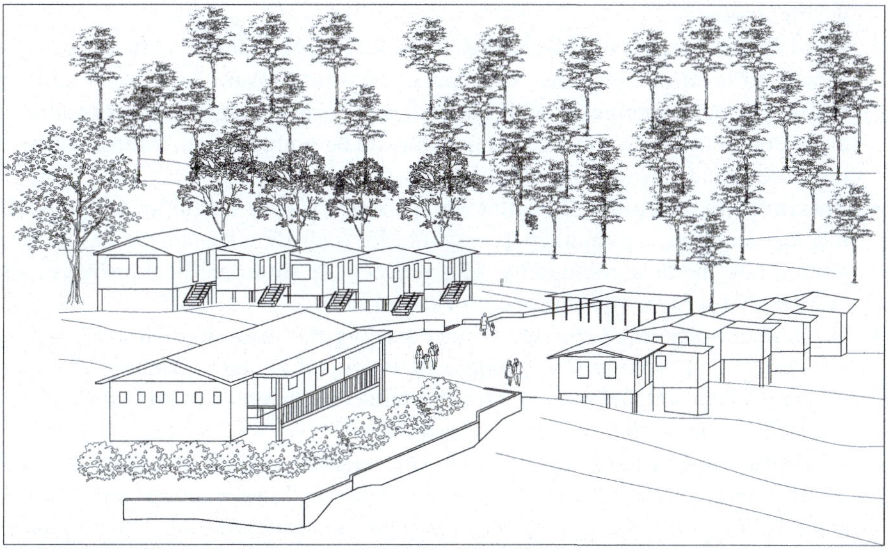

Fig. 2.8 Axonometric view of Tukuraki Village

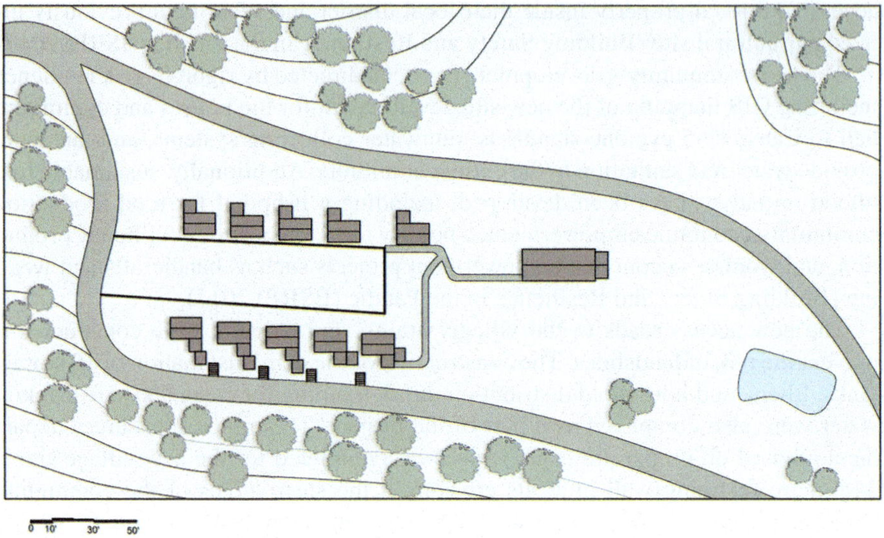

Fig. 2.9 Plan of Tukuraki Village

2.9.1 Lessons Learned

- **Adequate space allocation**: Overcrowding in refugee camps increases morbidity and stress. Therefore, providing sufficient space is crucial for both shelters and outdoor areas. The UNHCR recommends 45 sq. m. per person, including gardening space. Ensuring this standard promotes better living conditions and health outcomes.
- **Land use and legal considerations**: Land for refugee camps is often scarce and subject to legal complexities, especially regarding private land and traditional rights. Formal agreements with national and local authorities are essential to avoid conflicts over land use. Refugees should be granted clear, limited access to resources like firewood and grazing areas.
- **Proximity to amenities**: Locating camps near existing amenities like schools, medical facilities, and markets is essential for long-term sustainability and integration into the host community. This reduces the strain on local resources and ensures that refugees have access to vital services.
- **Natural conditions of the site**: Understanding the natural conditions, such as soil, slope, and proximity to floodplains, is critical in camp planning. Poor site selection can lead to additional infrastructure costs, maintenance challenges, and increased vulnerability to natural disasters.
- **Flexibility and future growth**: Camp sites must be adaptable to future growth, as the initial number of refugees may increase. Infrastructure and facilities should allow for modifications and expansions to accommodate changing needs, ensuring long-term sustainability and functionality.

- **Environmental and climatic considerations**: Refugee camps should be designed to fit the natural environment, leveraging the site's natural advantages while avoiding areas prone to flooding, soil instability, or environmental hazards. This minimizes infrastructure costs and maximizes safety.
- **Health and safety**: Site selection must take into account health risks, such as the presence of diseases, polluted areas, or seasonal variations in climate. Settling refugees in areas with drastically different environmental conditions can have adverse health impacts.
- **Accessibility**: Camps must be accessible year-round via reliable road networks for the distribution of relief supplies and other essential services. Roads and pathways within the camp should cater to all, including those with mobility challenges, ensuring inclusivity and functionality.

Weblink of Case Study Project: https://disasterdisplacement.org/wp-content/uploads/2022/12/Information-Paper_Tukuraki_compressed-1.pdf

2.10 Conclusion

Effective planning and site selection for refugee camps are crucial to ensuring the health, safety, and well-being of displaced populations. Selecting a site with an appropriate location, orientation, and landscape can contribute to the energy conservation of each dwelling. The site's proximity to existing infrastructure will lead to substantial savings as the cost of roads and essential utilities can be avoided. Sites with 5–10% slopes will be more suitable for small, detached homes due to the planning flexibility they provide. Following the site selection criteria can solve many problems that might arise during site planning, construction, and habitation process. Additionally, ensuring accessibility to services and infrastructure, while addressing health and environmental risks, promotes both immediate relief and long-term resilience. A well-planned refugee camp provides shelter and supports social cohesion, security, and integration into the local community.

Discussion Questions

- Why is it crucial to select well-planned sites for emergency shelters from the early stages of a refugee emergency?
- What factors should be considered to ensure the safety, security, and protection of displaced people when selecting a site for emergency shelters?
- How can local environmental issues impact the site selection process for refugee settlements?
- What are the potential long-term environmental impacts of poorly planned refugee camps, and how can they be minimized?
- Why is it important to consider the long-term availability of services when selecting a site for emergency shelters, even in temporary situations?
- What are the disadvantages of unplanned, improvised camps, and how can these be addressed through better site selection?

References

LaGro JA Jr (2007) Site analysis: a contextual approach to sustainable land planning and site design, 2nd edn. Wiley, Hoboken, NJ

Ahmed MM, Sulaiman M (2003) Design and proper sizing of solar energy schemes for electricity production in Malaysia. In: National Power Engineering Conference, PECon 2003—Proceedings. Institute of Electrical and Electronics Engineers Inc., pp 268–271

Atkinson W (2018) The orientation of buildings: or, planning for sunlight. Forgotten Books, London

Building Safety & Resilience in the Pacific (BSRP) (2017) Tukuraki village relocation project. Building safety & resilience in the Pacific. https://bsrp.gsd.spc.int/wp-content/uploads/2016/09/Tukuraki-Project-Brief.pdf. Accessed 13 Aug 2021

Duncan CI (1992) Soils and foundations for architects and engineers. Springer, New York, NY

De Chiara J, Koppelman L (1975) Urban planning and design criteria, 2nd edn. Van Nostrand Reinhold, New York, NY

Cuny FC (1977) Refugee camps and camp planning: the state of the art. Disasters 1(2):125–143. https://doi.org/10.1111/j.1467-7717.1977.tb00020.x

De Sousa C (2021) Sustainable brownfield development: building a sustainable future on sites of our polluting past. Taylor & Francis Ltd., London

Geoscience E and MD (n.d.) Refusing to disappear: How the people of Tukuraki became symbols of a Resilient Pacific. https://gem.spc.int/news/2021/06/refusing-to-disappear-how-the-people-of-tukuraki-became-symbols-of-a-resilient-pacific. Accessed 13 Aug 2021

Gilmer M (2019) Living on flood plains and wetlands: a homeowner's handbook. Lyons Press, Guilford, CT

Holden R, Liversedge J (2011) Construction for landscape architecture. Laurence King Publishing, London

International Organization for Migration (2019) Thousands displaced as monsoon rains pound rohingya camps. International Organization for Migration. https://www.iom.int/news/thousands-displaced-monsoon-rains-pound-rohingya-camps. Accessed 1 Mar 2024

Numbers MJ (1995) Sitting a house: proper sitting and design strategies enhance energy efficiency. Fine Homebuilding, pp 40–45, February/March, 1995SPC

Sphere (2018) The sphere handbook humanitarian charter and minimum standards in humanitarian response. Switzerland, Geneva

United Nations High Commissioner for Refugees (2015) Handbook for emergencies, 4th edn, Geneva. https://emergency.unhcr.org/. Accessed 1 June, 2024

United Nations High Commissioner for Refugees (2007) Handbook for emergencies. United Nations High Commissioner for Refugees, Geneva

Wietek B (2022) Slopes and excavations, 1st edn. Springer, Wiesbaden

Yerro I (2011) Greenery in architecture: vegetation in innovative European multifamily houses. VDM Verlag Dr, Müller

Chapter 3
Site Planning

Keywords Sustainable site planning · Refugee housing · Refugee camp planning · Refugee health and well-being · Long-term refugee needs · Topography and location of shelter · Open spaces integration in camp planning · Ecological footprint for refugee camp · Modular site planning for refugee camp · Rohingya refugee housing

The site planning of a camp significantly impacts the sustainability, health, and well-being of the refugee community. Sustainable planning must consider the long-term needs of the occupants, and therefore, planners and designers need to follow sustainable housing master planning principles early on in the process. The site's general layout, the orientation of structures, and its proximity to natural and existing features should be planned considering the topography, location, and passive solar gain and cooling. Additionally, the inclusion and seamless integration of open spaces and mobility networks are vital to the functioning of the place.

This chapter sets indexes for sustainable site planning and describes the necessary factors to be considered for long-term stay in refugee housing. It emphasizes the need for macro thinking when considering the form and density of housing, given their significant impact on ecological footprint and carbon emissions. The chapter highlights the importance of contingency planning in gathering demographic and environmental data to identify risks and inform site development. Modular site planning with mixed land use is endorsed for future growth and adaptability. The subsequent sections delve into density considerations, analyzing low-, medium-, and high-density categories, each with distinct advantages and challenges. Environmental considerations highlight the need to integrate site planning with environmental preservation, addressing grading, topography, and local resources. The chapter underscores the importance of maintaining the site's ecosystem and adapting to natural topography. Lastly, accessibility, mobility, and open spaces are discussed, emphasizing designing road networks, connecting communities, and integrating green spaces. A case study on the Rohingya refugee transitional housing in Bhasan Char, Bangladesh, demonstrates the practical implementation of these concepts, showcasing how thoughtful site planning can provide sustainable

A. Friedman, B. Chaki, *Fundamentals of Planning and Designing Sustainable Post-Disaster Shelters*, https://doi.org/10.1007/978-3-031-83317-5_3

solutions for displaced populations, considering factors like location, construction, social well-being, and economic empowerment.

3.1 Form and Density: The Macro Thinking

The form and density of refugee housing will significantly impact its sustainability, carbon, and ecological footprint. Therefore, contingency planning can help gather information regarding the site's environmental and demographic information about the occupants and identify possible risks (United Nations High Commissioner for Refugees 2007). Modular site planning with mixed land use benefits refugees' sustainable living environment and future expansion. This section discusses aspects related to the form of a place.

3.1.1 Density

The density of a refugee community will determine a multitude of planning considerations. Generally, *residential density* refers to the number of inhabitants or dwellings per unit area (Friedman 2012). This form of density, which is consistently lower than *net density*, identifies the relationship between the built mass and open space, often public areas such as parks, which is a good indication of the quality of life the residents will have. The refugee camps can be categorized as low, medium, or high density, each with unique advantages and disadvantages.

A low-density neighbourhood comprises large amounts of land, resulting in increased infrastructure and road costs. In medium-density projects, a pedestrian network can be introduced because amenities are closer, allowing for feasible mobility. Moreover, infrastructure costs are reduced, and some land preserved as public space can become an extension of private outdoor areas. Density at its highest level can lead to overcrowding with many negative ramifications. The camp's design must maximize the density within a small area without causing social challenges. Instead of merely supplying emergency shelter, the camp's development should result in sustainable settlements (Cuny 1977). Alshoubaki and Zazzara (2020) showed by analyzing the fragility indicators (i.e. socio-political fragility, economic fragility, education, and health system fragility) in Jordan refugee camps that 73% of the camps were fragile due to the camp area and 66.6% of the camps are fragile for high population density (Alshoubaki and Zazzara 2020). The refugee camps in Jordan have unprecedented densities (116 people per square kilometre), with ratios higher than the densities of the world's most populated cities. The rising number of refugees in the camps due to the growing birth rate has put their future in jeopardy (de Rooij et al. 2016).

Density is a critical factor in achieving a sustainable refugee neighbourhood. A refugee camp development can achieve sustainability by reaching the standard range of densities and incorporating open spaces following existing development around the site, as well as the various needs of the development. The appropriate density for refugee camps will largely depend on the context in which it is proposed. In many cases, urban development planning can mediate between low and high. Public outdoor spaces are integral to high-quality urban living as they are used for recreation, social integration, and physical activity. Smaller pockets of open space are more appropriate in higher-density planning as they are safer and less intimidating than less private, vast open spaces.

3.1.2 Environmental Consideration

Following an emergency, environmental concerns should be incorporated into site planning and shelter design from the outset (United Nations High Commissioner for Refugees 2007). In general, any development involves intervention and a change in the natural form of the land. Grading a site refers to the physical adjustment to this topography. It should be done with sympathy for the natural setting, particularly in cases where the development site is on hilly or sloped land (Holden and Liversedge 2011). The location and configuration of the camps, emergency shelter provisions, and the use of local resources for building and fuel may all significantly affect the climate. The most tremendous environmental damage generally occurs in the early stages of an emergency. Environmental degradation for the refugees and local population has health, social, economic, and political ramifications (United Nations High Commissioner for Refugees 2007).

A site's original topography is a key element in the local ecosystem, holding its distinctive flora and fauna. Therefore, maintaining the topography of a site is crucial in preserving the ecosystem. To maintain and adapt to the existing topography, roads and buildings should be sited with regard to slopes and changes in elevation and must follow contour patterns. Steep slopes should also be avoided for ease of construction. Working with the existing terrain and natural slopes reduces construction costs. Also, it leads to infrastructure that contains natural bumps and bends, leading to more cautious driving, increasing road safety for residents.

An effective way of arranging shelters on a sloped site is by using a terracing technique (Fig. 3.1). By arranging shelters in varying rows along the elevation, footholds—or a base for a solid foundation—can be created. By arranging buildings along the slope in a terraced sequence, views from each structure will be unobscured. Any space left between succeeding rows allows room for private outdoor space, decreases the obstruction of views, and prevents shadows of one building from another. A sloped terrain also allows for a simple, natural, and adequate drainage system, reducing costs.

Generally, a building should adhere to the site's existing natural conditions and fit with any existing buildings or infrastructure. Finding a suitable site by considering all environmental factors relating to elevation and topography also means

Fig. 3.1 Site plan showing shelters arranged in varying rows along the contour lines

considering the effects and potential gains from weather conditions, such as wind, light, or temperature. Chapter 2 explores the intricate details of site selection criteria.

3.1.3 Social Factors Consideration

When master planning the design and site layout for refugee housing, it is essential to consider various social aspects to ensure the development is sustainable, inclusive, and conducive to the well-being of its residents. These social aspects include socioeconomic, socio-environmental, and socio-cultural factors, each playing a critical role in the success of the housing project.

Socioeconomic factors are paramount in the planning process. Understanding the economic conditions of the refugee population is crucial for designing housing that is affordable and accessible. This includes considering the income levels, employment opportunities, and economic activities within the community. Providing access to vocational training and job opportunities can help refugees become self-reliant and economically integrated into the host community(Eichner and Ivanova 2018). Additionally, the cost of utilities and maintenance should be kept within the financial reach of the residents to ensure long-term sustainability.

Socio-environmental factors involve the interaction between the community and its environment. It is important to assess the environmental impact of the housing project and incorporate sustainable practices. This includes using eco-friendly construction materials, implementing energy-efficient systems, and ensuring proper waste management. The site layout should also consider the availability of natural resources and potential environmental risks, such as flooding or landslides, to ensure the safety and resilience of the housing (Eichner and Ivanova 2018). Integrating green spaces and recreational areas within the site plan can enhance the quality of life by providing residents with areas for relaxation and social interaction (UN-Habitat and UNHCR 2020).

Socio-cultural factors are critical in ensuring that the housing design meets the cultural and social needs of the refugee population. This includes considering the cultural diversity, traditions, and social norms of the residents. For instance, the design should accommodate different family structures and provide communal spaces that encourage social cohesion and community building. Educational and healthcare facilities should be culturally sensitive and equipped to address the specific needs of the refugee population. Community participation in the planning process is vital to ensure that the housing design reflects the actual needs and preferences of the residents.

In conclusion, considering socioeconomic, socio-environmental, and socio-cultural factors in the master planning design of refugee housing is essential for creating sustainable and inclusive communities. By addressing these social aspects, planners can ensure that the housing not only meets the immediate needs of the refugees but also contributes to their long-term well-being and integration into society.

3.2 Planning Typologies

Standard camp plans are categorized based on the main limiting factor determining their selection. In these cases, the terrain or the type of housing is the limiting factor. There are three classifications of camps: modular planning, grid planning, and cluster planning. These typologies are elaborated below.

3.2.1 Modular Site Planning

Modular camp planning employs a module repetitively in the site master plan design to establish the sizes of shelters, community buildings, open spaces, pavements, and block patterns. In Fig. 3.2, a module of 1219 mm by 1219 mm (4 ft. by 4 ft.) determines the site's shelter sizes, community buildings, roads, and open spaces. The module size can be increased and decreased to get optimum sizes for the site plan elements (i.e. shelters, intermediate spaces between shelters, and facilities). Adaptability and flexibility to future needs are vital for a long-term stay in

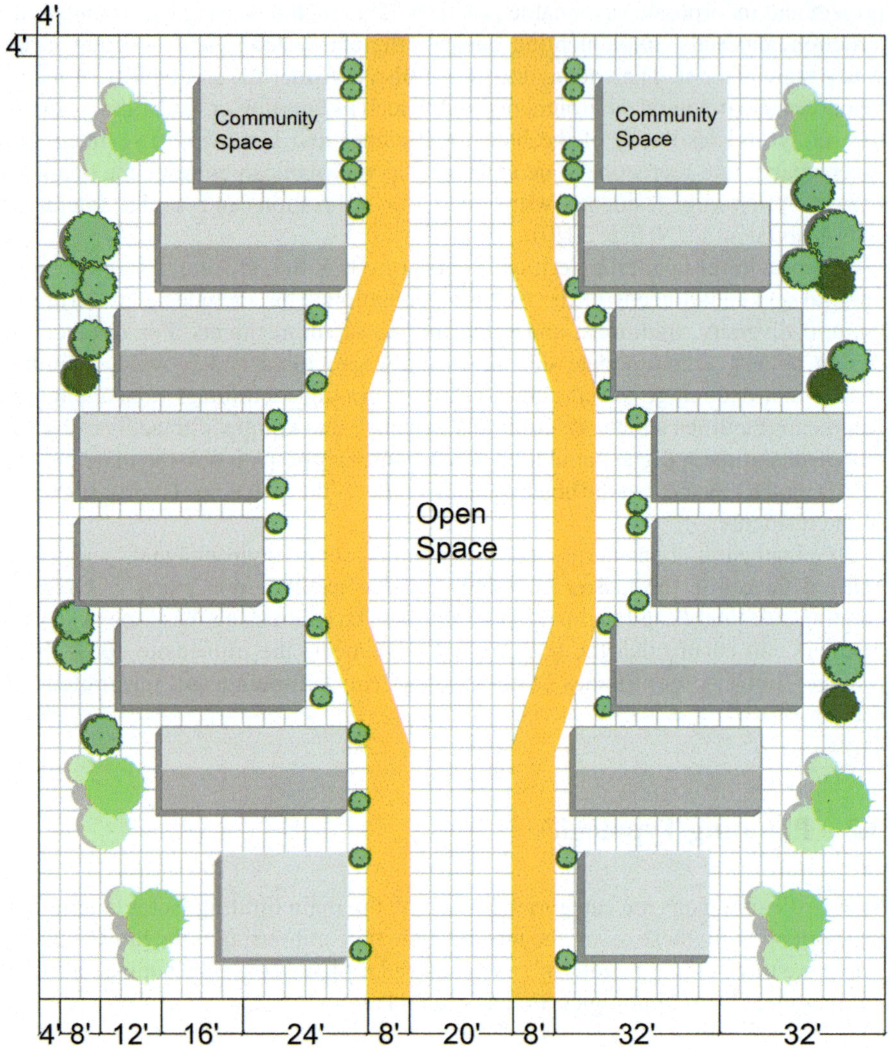

Fig. 3.2 Modular camp master planning

emergency shelters. Hence, modular planning can be used as it allows future adaptability and minimizes material waste. Modular site planning should begin with a refugee household by considering the family's needs, such as distance to water and latrines, as well as conventional housing and living arrangements. The site planning and physical organization should begin with the smallest module, the family, and move to larger units (United Nations High Commissioner for Refugees 2007). Table 3.1 shows the recommended size of a module according to UNHCR.

Table 3.1 Recommended module size. Source: (United Nations High Commissioner for Refugees 2007)

Module	Consist of	Approximate number of persons
Family	1 family	4–6 persons
Community	16 families	80 persons
Block	16 communities	1250 persons
Sector	4 blocks	5000 persons
Camp module	4 sectors	20,000 persons

3.2.2 Grid Planning

Placing roads, shelters, and administrative and communal facilities in a grid has been a common approach to past site planning (Corsellis and Vitale 2005).The linear pattern, which consists of square or rectangular areas separated by parallel streets, is commonly used due to its ease of design and implementation (United Nations High Commissioner for Refugees 2007). This planning typology requires the least space to demarcate and divide individual areas. This layout is often the most widely used to achieve high density because of the compactness and standard size of the different land uses.

A compact refugee camp facilitates the spread of disease. It can promote undesirable social conditions in the camp, increasing administrative and control costs in the long run. Therefore, grid planning is the least desirable method due to an attempt to achieve high density. While the grid plan is better suited to flat, even terrain, it can be tweaked to operate in areas with minor elevations. The grid should not be used in hilly or mountainous terrain because the drainage and sewage systems would not work correctly. Grid camps resemble military camps in appearance and are often designed by military engineers using "ordinary" military site-planning plans. This type of planning is developed for military camp design and is not recommended for a refugee camp. This is because the needs of a temporarily deployed military camp differ from those of multiple generations of refugees in a semi-permanent settlement. The military can be transferred or redeployed depending on their needs, whereas refugees stay in refugee camps for several years. Therefore, grid planning might not meet essential criteria for refugee housing, such as adequate space for family living (Cuny 1977). If the grid planning layout is selected, the master plan should include sufficient infrastructure and amenities to maintain living conditions in the refugee neighbourhood for the long term.

3.2.3 Cluster Planning

Cluster planning refers to when main road networks are set up in the shape of a tree, with roadways spreading from central areas that serve communal amenities (Fig. 3.3). To build a hierarchy of road infrastructure rather than an undifferentiated

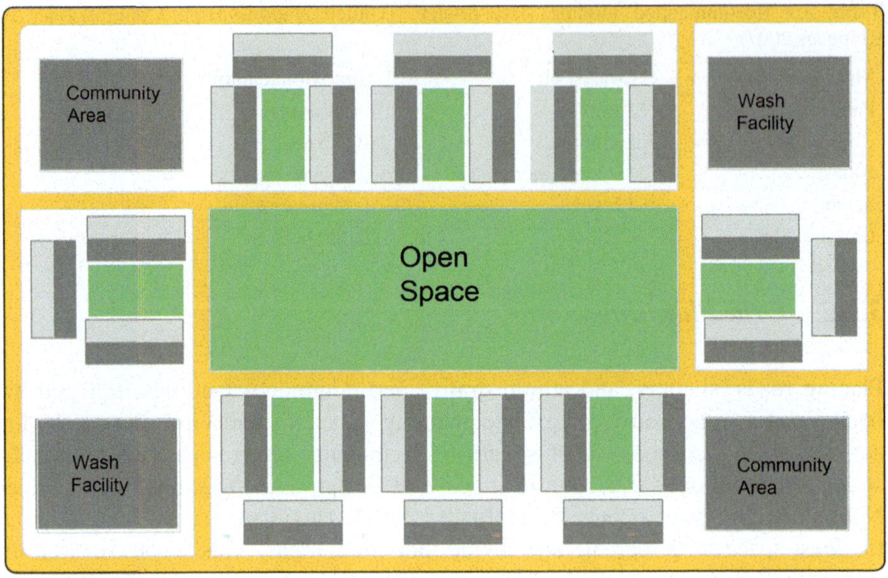

Fig. 3.3 Cluster camp master planning

grid, road sizes can be varied depending on the volume of use in each case. Cluster planning is advantageous to grid planning because it can improve communities by developing "private" areas and promoting communal activities and practices, such as water collection and cooking. Furthermore, cluster planning encourages social hierarchy, which can help accept services and include displaced people's needs (Corsellis and Vitale 2005). The case study of Rohingya refugee transitional housing (Sect. 3.6) is a great example where the housing design utilized the open space for agriculture and a pond for fish farming, giving refugees new opportunities to make money and engage in constructive activities that improve their general well-being and self-reliance.

3.3 Mixed Land Use

Refugee neighbourhoods with mixed land use and various dwelling types help limit sprawl. They also offer richer built environments and social opportunities to their residents. Mixed communities offer these groups stimulating, active, convenient, and social lives. Living in a diverse neighbourhood with civic, community, and commercial institutions nearby benefits all residents. Moreover, commercial activities inside the refugee neighbourhood can inject money into the affected community.

Mixed-use planning and increasing density can create more open spaces (Hall 2007). Varying land uses allow for community gardens, parks, and other green spaces to be integrated alongside residential and commercial developments,

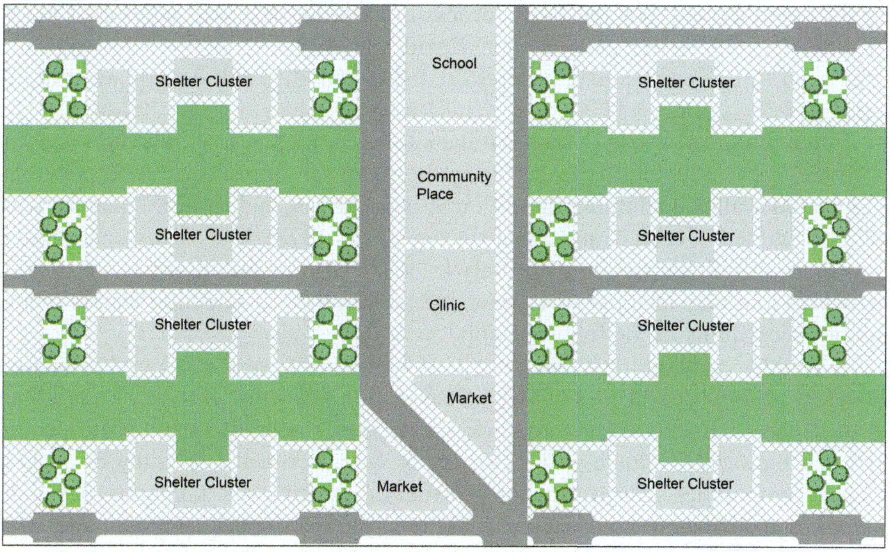

Fig. 3.4 Mixed land use with amenities for long-term needs

allowing even distribution. Green spaces are of great ecological, social, and economic importance. They disrupt the urban heat island effect, encourage an active lifestyle, preserve biodiversity, and improve food security. Diverse communities benefit from shared outdoor space through an increased sense of cohesion. Integrating public open spaces amidst different dwelling types will encourage people to interact with their neighbours and engage with their surroundings more meaningfully. An exciting mix of residential, office, commercial, and entertainment spaces also draws visitors from outside the neighbourhood as well (Grant 2002) (Fig. 3.4).

Mixed-use planning methods can be categorized based on neighbourhood configuration, spatial organization, transit nodes, and retail hubs. When planning for mixed dwelling types, designers must consider all aspects of a neighbourhood such as the demographic criteria, the built environment, and the aesthetics of the surrounding areas. A well-designed mixed-use neighbourhood will encourage its residents to walk, take transit, use community organizations, and participate in the local economy. "Neighbourhood configuration" refers to the design and connectivity of different landscape elements. Considering neighbourhood configuration can optimize the benefits of mixed-use planning (Hersperger 2006).

3.4 Active Mobility

Well-connected communities are also beneficial to residents' health. Designing mobility and connectivity networks with suitable integration of streets, bike, and pedestrian paths and reducing the use of private vehicles are central to planning smart communities. For the most successful creation of a well-connected neighbourhood with lower levels of carbon dioxide emissions, some basic design principles need to be considered. Concepts related to mobility will be explained in this section.

Camp planners should ensure that the site has sufficient road infrastructure and safe and easy accessibility, even during the rainy season. Internal road networks, significant walkways linking the different usage areas, and local footpaths should all be included in each camp's road system (Cuny 1977). The road would provide refugees with direct access to not only trade but also to any informal information networks that may take place. Chapter 2 describes the site selection factors related to the accessibility of the refugee camp.

A healthy community is one where everyone has access to a good education, secure housing, decent work, and transportation. Healthy or typical streets are community-wide networks of roadways and connector trails intended for people, not just motorized vehicles. Road hierarchy is less critical in healthy or conventional communities. They have narrow avenues, short blocks, numerous ties, sidewalks, and deliberate landscaping. Children can also walk to schools and other destinations independently because of the layout, accessibility, path options, and strong support for cycling, which decreases the number of regular car trips (Burden 2000).

It is recommended that time be spent designing trails, alleys, lanes, streets, main streets, boulevards, and parkways for specific use to shape healthy streets. The streets are classified according to the services they provide to the community. Street categories can be divided into the following three types (Burden 2000) (Fig. 3.5):

- **Category 1**: Tertiary lanes such as trails, alleys, and streets that provide neighbourhood access.
- **Category 2**: Secondary roadways such as avenues and main streets that provide access to neighbourhood streets.
- **Category 3**: Primary roadways such as boulevards and parkways that provide regional access.

Sustainable street networks connect people's homes, workplaces, schools, and shopping areas. Pedestrians, bicyclists, and wheelchair users should be able to choose from several routes to reach their destinations. After the construction process is completed, the road that runs through the refugee neighbourhood provides more trade opportunities for the refugees, particularly once their livelihoods have been established to the point where some of them are producing products for export out of the camp as well as importing items for sale within the camp.

People of all ages populate safe streets and participate in a variety of activities, such as eating at sidewalk cafés, shopping, relaxing on benches, chatting with friends, reading newspapers, window shopping, strolling, jogging, meeting new people, and walking dogs—as many as can be imagined. Thus, sustainable street networks can remove barriers between the refugee and host communities, facilitating more harmonious relations between different groups within and between both communities (Burden 2000).

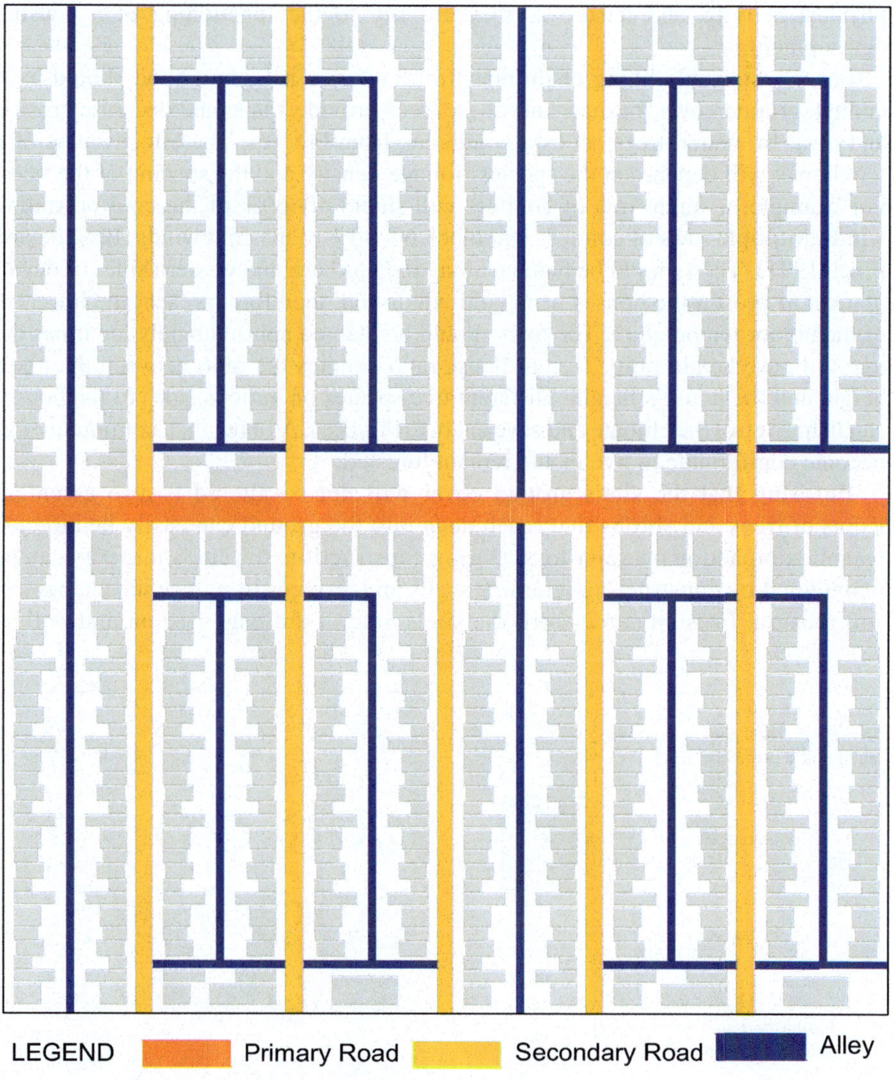

Fig. 3.5 Street categories

3.5 Open Spaces and Natural Elements

Though traditional refugee housing tends to focus primarily on building areas provided for the construction of the shelters, the creation of open spaces is equally significant for their long-term stay. According to UNHCR, 12 to 20% of the site area should be open space (United Nations High Commissioner for Refugees 2015). Within this context, the term "open spaces" refers to areas left unbuilt while adjacent sectors are developed. Refugees rely on open spaces for mental health, physical activity, and leisure as they are key to the functioning of a housing development.

For a sustainable living environment, the refugee neighbourhood should have enough ground cover, bushes, and trees. Trees provide shade and decrease erosion and dust. In winter, trees act as a windbreak from breezes. The effectiveness of a windbreak will depend on the species' foliage density and the spacing of the trees. For example, a single row of conifers will shield 40–60% of the cold prevailing winds. Multiple rows of conifers will block 60–80% of the cold winds (Brandle and Finch 1991). They should be planted along the northern and western sides of development to protect the houses from cold winds and, therefore, prevent the lowering of the indoor temperature. The trees should be planted approximately ten times the tree-to-house height ratio. Tree-to-house ratio denotes the ratio between the tree's height and the house's height. Different tree species on various sides of the houses can help control the climate and save energy (Fig. 3.6). As a result, a community can become comfortable to live in and remain forested.

Ensuring that the site's ground cover will not be altered is also essential. Roots of trees and shrubs that reach deep into the ground can easily be affected by soil compaction or water loss. During construction, the site's integrity can be maintained by constructing retaining walls and terraces to reduce the chances of soil erosion. Erosion can disturb the site's natural drainage system and pollute

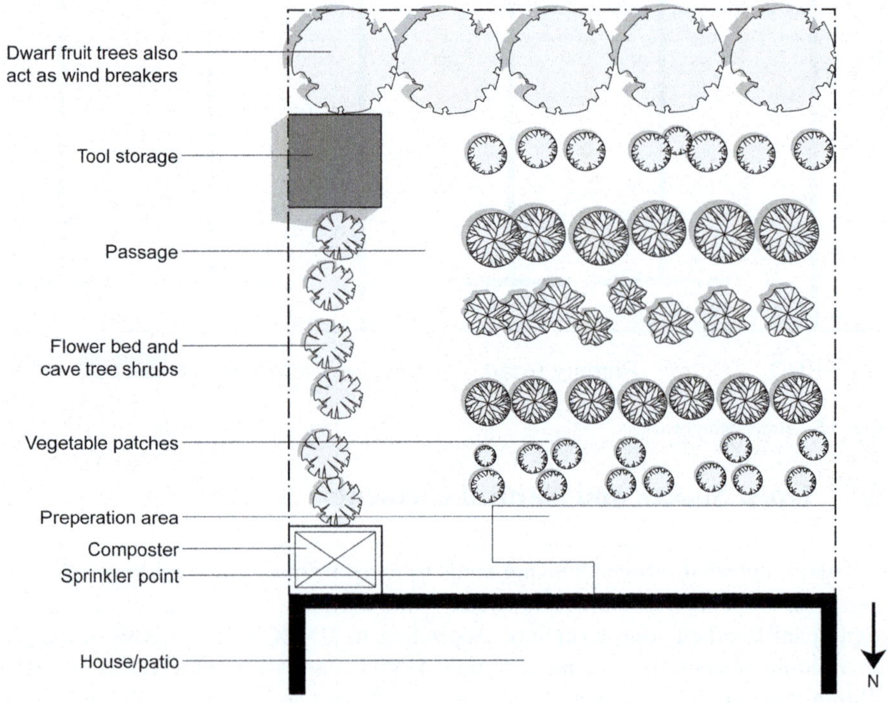

Fig. 3.6 Different species of tree plantation for different benefits

nearby water sources (Friedman 2018). Care should be taken during site planning to do as little damage to vegetation and topsoil as possible. Indiscriminate bulldozing or topsoil removal must be avoided at all costs when using heavy machinery. Suppose wood is required for domestic cooking or shelter building. In that case, the refugees should be alerted not to take their wood supplies from the camp's immediate vicinity (United Nations High Commissioner for Refugees 2015).

Maintaining the site's natural cover is also essential. As trees take a long time to grow, existing trees, water bodies, and plants on the site should be preserved as much as possible. Smaller plants or bushes can be cleared to make room for tents, other structures, and roadways. Large trees generally have the most expansive systems of roots to serve as soil binders and anti-erosion devices and help keep nearby indoor environments cool. If any trees need to be cut down, it should be done in a considered "triage" manner, with trees within shelter communities being removed first and trees lining or near roadways being removed last. The informal community meeting points, market stalls, alleys, individual shelters, and large public spaces can benefit greatly from trees' shade at the edges of major roadways in the camp or at the edge of public squares (Scott-Smith and Breeze 2020).

Natural open spaces and important features within the site should also be considered. The initial site plan must contain an analysis of existing and future needs, and it should retain any natural open space for encouraging recreational activity. To further benefit the community, planners of green open spaces should consider complementary sets of parks and open spaces that meet the needs of a full range of community interests that work for the majority of people. Large open spaces should be linked by pedestrian paths to major activity centres such as schools and employment centres to encourage physical activity through green spaces (Fig. 3.7). One innovative and sustainable idea to incorporate into the green open space for all residents' benefit is a garden. Most importantly, well-designed public spaces and landscapes allow for what Gehl has referred to as "life between buildings", which alludes to the spectrum of recreational, social, and cultural activities in public spaces, creating meaningful and attractive communal spaces (Gehl 2024).

In the last decade, it has become clear that human activity has damaged the natural integrity of ecosystems, threatening the security of the societies that depend on these systems (Roseland 2000). Passive solar gain, wind direction, preservation of flora, and topography are significant factors in a community's sustainability (Friedman 2018). Recognizing the implications of setting decisions early in the design process and ensuring the successful integration of natural and built environments will lead to the building of sustainable living for the refugees.

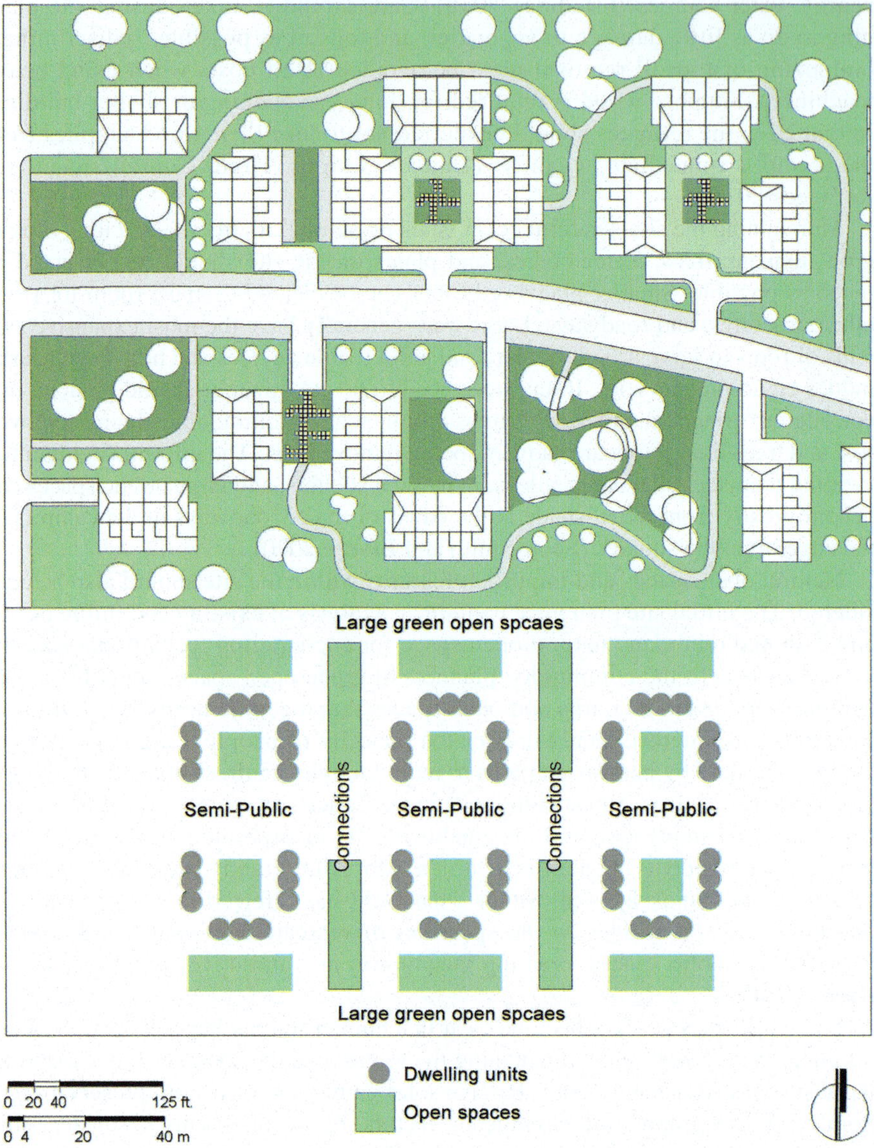

Fig. 3.7 The connection between open space and semi-public facilities

3.6 Case Study: Rohingya Refugee Transitional Housing, Bhasan Char, Bangladesh

The 34 refugee camps in Bangladesh's Cox's Bazar area are presently home to more than a million Rohingya people who were forcibly uprooted from their homes in Myanmar. In 2022, the government of Bangladesh began the process of moving some Rohingyas to Bhasan Char Island, which is situated in the Noakhali region of Bangladesh, to relieve the burden of Rohingyas in Cox's Bazar. This is a transitional housing project where the Rohingya refugees will live for the next few years.

Site Location

Bhasan Char, a Noakhali island with abundant natural beauty and sources of income, is prepared to house impoverished people with all the amenities. The island, about 21 nautical miles from the Noakhali district administrative center, has communications infrastructure for the general public with motorboats. At the same time, scheduled ships for commercial usage depart from Chattogram. Bhasan Char has a total area of 5261 hectares (13,000 acres), including low-lying regions, of which 2601 hectares (6427 acres) are now usable. Out of the available area, 174.8 (432 acres) were used to build the residential infrastructure, 688.8 hectares (1702 acres) were used to construct embankments, and 371.5 hectares (918 acres) were used to plant trees (Banerjee 2020).

Site Planning and Construction

As part of the Ashrayan-3 project, the Bangladeshi government assigned the navy to transform Bhasan Char so it could be inhabited. About 100,000 refugees will be housed in semi-permanent shelters, which will eventually be able to handle 400,000. On an area of 6.7 square kilometres (2.5 square miles), rows of identical steel and concrete buildings have been constructed around a central courtyard with a pond. There are 16 spacious rooms in each building, and each one can hold up to four people. Two communal kitchens are used by the entire block, along with six restrooms and four bathrooms for men and women.

Each cluster includes a multifunctional shelter that can house 200 heads of cattle and approximately 1000 people and resist winds up to 260 km (161 miles) per hour. Additionally, the shelters may be utilized for administrative, educational, medical, and other neighbourhood-related activities. According to reports, two of the 120 shelters will serve as hospitals for the UN delegates, the Refugee Relief and Repatriation Commissioner, law enforcement officials, and each of those groups' representatives. One of the buildings will serve as a "megastore", three as mosques, four as community clinics, and one as a homeless shelter. Moreover, guest homes, a daycare center, and an orphanage have been constructed.

Along with facilities for solar electricity, solar pumps, and biogas, provisions for rainwater collection have been developed. The water will be delivered to the clusters by solar pumps from deep tube wells dug to a depth of 213–228 m (700–750 ft.). Access roads to the landing sites, crossroads, and walkways inside the clusters have all been built. Four warehouses are being constructed to contain both food and non-food items. A jetty, a dam, a fire station, and a weather bureau will also be on the island. Although a police camp has already been established on Bhasan Char, installing 120 cameras, building watchtowers, and constructing a surrounding fence will improve security and monitoring.

Bhasan Char is, on average, 2.84 m (9.31 ft.) above mean sea level. Thus, many low-lying areas have been filled with sand to increase their height and shield the new center from flooding and river erosion. A three-metre-high flood defense barrier would also shield the island from tidal surges and other natural calamities. Eight high-speed boats and two helipads have been built for emergency transportation and evacuation (Banerjee 2020) (Figs. 3.8 and 3.9).

Social and Economic Freedom

The region is filled with greenery. Paddy and other vegetables might be grown using an organic approach in addition to cleansed water from rivers and other regions. They rely primarily on harvesting outputs. There is a lot of unclaimed land available for crop cultivation. Moreover, Bhasan Char is surrounded by rivers and water sources 10 ft. deep, such as ponds, streams, and large lakes.

A sizable portion of the Rohingya population was employed directly or indirectly in the fish industry. Additionally, the area around the large pond has plenty of room for planting crops and vegetables. This acreage makes it simple for people to cultivate their food and meet their needs. Fish farming also has a lot of potential (Islam et al. 2022).

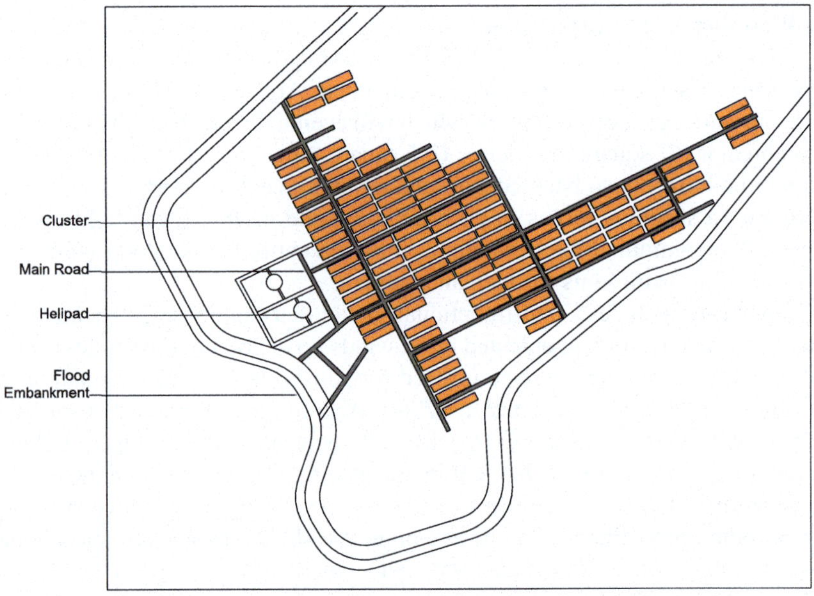

Fig. 3.8 Masterplan of Rohingya refugee transitional housing

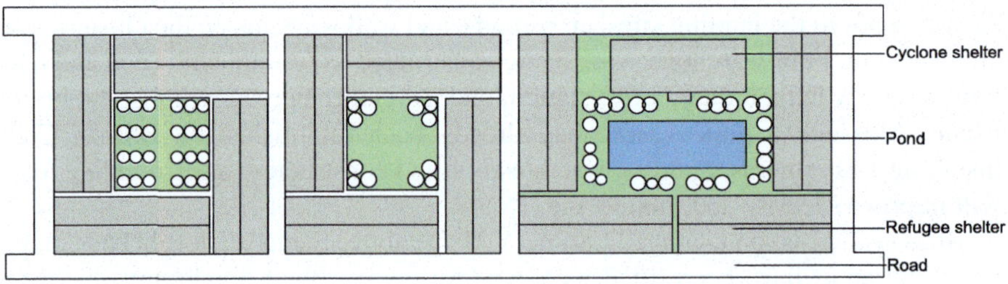

Fig. 3.9 Cluster plan of Rohingya refugee transitional housing

3.6.1 Lessons Learned

- **Sustainable site planning**: Proper planning of a refugee camp's layout, structure orientation, and integration with natural features ensures sustainability and long-term viability for the community. Incorporating factors such as topography, passive solar gain, and cooling from the outset minimizes environmental impacts and improves the quality of life for residents.
- **Form and density considerations**: The density of a camp significantly affects its sustainability and livability. Low-density camps require more infrastructure, while medium-density designs balance mobility and public space. High-density planning should avoid overcrowding to prevent social issues, focusing on sustainable development instead of emergency shelter provision.
- **Environmental considerations**: Site planning must integrate with the natural environment by preserving the ecosystem, respecting topography, and ensuring proper drainage and grading. Working with natural slopes reduces construction costs and improves road safety, while terracing can optimize space usage and views.
- **Social factors**: Sustainable refugee housing must consider socioeconomic, environmental, and cultural factors. This includes ensuring access to employment opportunities, incorporating green spaces, and designing communal areas that reflect the cultural and social needs of the population.
- **Planning typologies**: Different planning models—modular, grid, and cluster—have varying benefits. Modular planning offers flexibility for future expansion, while grid planning is compact but less desirable due to potential overcrowding. Cluster planning encourages community interaction and shared amenities.
- **Mixed land use**: Mixing administrative, educational, medical, and communal spaces promotes a vibrant, integrated refugee neighbourhood. It allows for social interaction, economic opportunities, and the efficient use of space while improving the ecological and social well-being of the community.
- **Active mobility**: Well-connected road networks, paths for pedestrians and cyclists, and reduced reliance on private vehicles are crucial to promoting mobility within camps. Proper planning ensures residents have access to essential services and encourages healthier, more sustainable communities.
- **Open spaces and natural elements**: Incorporating open spaces and natural elements like trees, green areas, and water features is critical for the well-being of residents. These spaces provide recreational opportunities, improve mental health, and help preserve the local ecosystem.

Weblink of Case Study Project: https://fingfx.thomsonreuters.com/gfx/rngs/MYANMAR-ROHINGYA/010060Z21XP/index.html

3.7 Conclusion

To conclude, site planning is critical in achieving a sustainable and healthy built environment for the refugees. While planning, the topography, sun, and wind should be considered so that the benefits of the elements are maximized for energy efficiency and that homes and buildings fit seamlessly into the landscape. By mixing various land use and clustering amenities, the planners can support the needs of a socioeconomically and racially diverse group of residents while encouraging an active lifestyle. Furthermore, it is essential to utilize strategies that contribute to the interconnected nature of open space systems and variability in mobility and intergenerational usage. Lastly, master planning that includes an appropriate and comfortable sense of proportion leads to more sustainable communities for a long-term stay.

Discussion Questions

- What elements should be included in a "master plan" for refugee housing, and why are they important?
- How does the social organization of refugees influence the principles of camp design and physical features in site planning?
- What criteria should be considered for sustainable living conditions and long-term stays in refugee camps?
- How do natural features like site contour levels, flood plains, and soil types impact the master planning of refugee sites?
- Why is it important to collect environmental data during site planning, and how does it inform site development?

References

Alshoubaki H, Zazzara L (2020) The fragility in the land of refugees: Jordan and irrepressible phenomenon of refugee camps. J Int Stud 13(1):123–142. https://doi.org/10.14254/2071-8330.2020/13-1/8

Banerjee S (2020) From Cox's bazar to bhasan char: an assessment of Bangladesh's relocation plan for rohingya refugees. Observer Research Foundation

Brandle JR, Finch S (1991) How windbreaks work. Papers in Natural Resources 121. University of Nebraska—Lincoln

Burden D (2000) Street design guidelines for healthy neighborhoods. Transportation research circular. Walkable Communities, Inc., High Springs, FL

Corsellis T, Vitale A (2005) Transitional settlement, displaced populations. Oxfam GB University of Cambridge

Cuny FC (1977) Refugee camps and camp planning: the state of the art. Disasters 1(2):125–143. https://doi.org/10.1111/j.1467-7717.1977.tb00020.x

de Rooij LL, Wascher DM, Paulissen MPCP (2016) Sustainable design principles for refugee camps. Wageningen Environmental Research, Wageningen

Eichner M, Ivanova Z (2018) Sustainable and social quality of refugee housing architecture. MATEC Web of Conferences 193:04001. https://doi.org/10.1051/matecconf/201819304001

Friedman A (2012) Fundamentals of sustainable dwellings. Island Press/Center for Resource Economics, Washington, DC

Friedman A (2018) Smart homes and communities. Images Publishing Dist Ac, Mulgrave

Gehl J (2024) Life between buildings, 6th edn. Island Press, Washington, DC

Grant J (2002) Mixed use in theory and practice: Canadian experience with implementing a planning principle. J Am Plan Assoc 68(1):71–84. https://doi.org/10.1080/01944360208977192

Hall E (2007) Divide and sprawl, decline and fall: a comparative critique of Euclidean zoning. Univ. Pitt Law Rev. 68:915–952

Hersperger AM (2006) Spatial adjacencies and interactions: neighborhood mosaics for landscape ecological planning. Landsc Urban Plan 77(3):227–239. https://doi.org/10.1016/j.landurbplan.2005.02.009

Holden R, Liversedge J (2011) Construction for landscape architecture. Laurence King Publishing Ltd., London

Islam MM, Barman A, Khan MI, Goswami GG, Siddiqi B, Mukul SA (2022) Sustainable livelihood for displaced rohingyas and their resilience at bhashan char in Bangladesh. Sustain For 14(10):6374. https://doi.org/10.3390/su14106374

Roseland M (2000) Sustainable community development: integrating environmental, economic, and social objectives. Prog Plann 54(2):73–132. https://doi.org/10.1016/S0305-9006(00)00003-9

Scott-Smith T, Breeze ME (eds) (2020) Structures of protection? Rethinking refugee shelter, 1st edn. Berghahn Books, Oxford, NY

UN-Habitat, UNHCR (2020) Settlement profiling tool: a spatial analysis framework for settlements accommodating displaced populations

United Nations High Commissioner for Refugees (2007) Handbook for emergencies. United Nations High Commissioner for Refugees, Geneva

United Nations High Commissioner for Refugees (2015) Handbook for emergencies, 4th edn, Geneva

Chapter 4
Planning for Rapid Delivery

Keywords Refugee shelter design · Emergency preparedness · Emergency response · Refugee housing · Contingency planning for displacement · Rapid shelter delivery · Displaced populations · Parametric shelter design · Shelter design generation · Refugee camp planning · Sustainable refugee housing

Following conflict or environmental disaster, it is necessary to take immediate action to provide shelter for displaced people. Despite the difficulties of operating in the unexpected and ever-changing environment of a humanitarian crisis, well-planned programs with robust monitoring are constantly able to deliver within timetables and budgets. Planning and execution must start as soon as possible since responses to shelter, and settlements take time. Planning and coordination facilitated the building of refugee housing for 17,500 households in response to the large-scale displacement caused by military operations in Mosul, Iraq, in less than 6 months (Global Shelter Cluster 2021).

Planning helps to predict risks that might emerge during the construction process. For fast delivery, contingency planning can be helpful to avoid risks and reduce execution time. During contingency planning, it is essential to identify environmental issues and concerns that might be necessary to design shelters and their camps. Furthermore, with the advent of computational and parametric design technologies, there is an opportunity to revolutionize the rapid delivery of refugee housing. This chapter will outline the planning process for fast and quality delivery of emergency shelters. It will also explore how computational and parametric design approaches can facilitate the creation of efficient, sustainable, and scalable refugee housing solutions to meet the urgent needs of displaced populations.

A. Friedman, B. Chaki, *Fundamentals of Planning and Designing Sustainable Post-Disaster Shelters*, https://doi.org/10.1007/978-3-031-83317-5_4

4.1 Rapid Planning for Emergency Preparedness and Emergency Response

UNHCR defines planning as implementing the procedure for analysing the situation, identifying short-term and long-term targets, and the activities to achieve them. Both before and during a disaster, preparation is essential. Operations preparedness must be based on thorough needs, resources, and participation assessments with women, girls, boys, and men of concern. Women refugees often require shelter solutions that provide privacy and protection from the potential risks of gender-based violence. Gender-specific accommodations such as separate sleeping and sanitation facilities are crucial to ensure the safety and dignity of women. Young boys and girls need designated areas where they can play, learn, and receive psychosocial support to cope with the trauma of displacement. Children also require secure sleeping spaces, preferably separate from unrelated adults, to ensure their safety.

Establishing a multi-functional team to guarantee a comprehensive response, assigning and defining roles and duties for everyone engaged, structuring communication and information flow, and selecting, training, and managing personnel are all components of organizing and coordinating. One of the most critical aspects of planning in an emergency is coordination between UNHCR and outside parties (Humanitarian Response 2021).

Table 4.1 illustrates one possible model of activities that could take place in a refugee crisis. It is essential to realize that separate phases and actions of a refugee emergency operation may coexist or occur simultaneously. The last stage of an emergency operation is the transition from emergency response to longer-term

Table 4.1 Stages of emergency preparedness and response

Stage	Typical activities
Emergency preparedness	Prevention
	Early warning
	Contingency planning
	Development of emergency response systems
	Generation of support among potential host and donor governments
	Provision of stand-by resources
	Pre-positioning of supplies
	Training
Emergency response	Problem, needs, resources, and participatory assessments
	Community mobilization
	Resource mobilization
	Handling donor relations and media interest
	Operations planning
	Implementation and coordination
	Monitoring and evaluation
	Transition to the post-emergency operation

assistance, creating a community-based strategy and lasting solutions (e.g. voluntary repatriation, local integration, and relocation).

The amount of time spent providing emergency assistance should be kept to a minimum, and long-term considerations should always be factored into planning and execution. The need to balance the short and long term is evident in many crucial industries. The United Nations High Commissioner for Refugees (UNHCR) is required to help governments find long-lasting solutions to the refugee issue. The planners should always consider long-term solutions, beginning with the contingency planning phase. During this time, decisions are made on the type, quantity, and duration of help provided. How help is provided, and the roles played by the various community members may enhance or detract from their ability to fend for themselves. These decisions frequently impact the likelihood of long-lasting solutions that continue functioning even after the emergency ends (United Nations High Commissioner for Refugees 2007).

4.1.1 Emergency Preparedness

Being ready is the most excellent approach to guaranteeing a successful emergency response. Emergency preparedness can be defined as planning and taking action to ensure that the necessary resources will be available in time to fulfil anticipated disaster demands and that the appropriate infrastructure will be in place. Emergency preparedness has a broad range, and operations at that stage can be carried out at the international, regional, and national levels. Preparedness measures should enhance the emergency response capabilities of an organization.

Emergency preparedness can play a pivotal role in mitigating the impact of housing issues after disasters, and Cambodia serves as an illustrative case study. Situated in Southeast Asia, Cambodia faces numerous natural hazards, including floods, storms, typhoons, and cyclical droughts, which have profound economic and social consequences. Between 2018 and 2021, the government took proactive measures to address disaster-related concerns, particularly flooding and landslides. They established 48 participatory groups responsible for implementing a shelter and settlement improvement plan. This initiative aimed to bolster the disaster preparedness capabilities of vulnerable communities, thus reducing the potential adverse effects of disasters (International Organization for Migration (IOM) 2023).

UNHCR operates globally, maintaining a diverse range of available emergency response resources. These essential tools have been developed through insights from past emergencies, ensuring their effectiveness. The resources encompass operational support materials and services, personnel assistance, as well as human and financial resources, all centrally managed. Additionally, centrally controlled emergency stocks are in place. This setup enables swift deployment to any location requiring assistance, ensuring a consistent and predictable emergency readiness worldwide. Furthermore, UNHCR continually conducts training exercises to enhance its capacity and preparedness. UNHCR's emergency preparedness tools have also been employed in response to various natural disasters that have resulted in population displacement. For example, during hurricanes, earthquakes, floods, and other calamities, UNHCR and the Red Cross have provided aid and protection to displaced people in affected regions (UNHCR n.d.-b).

4.1.2 Emergency Response

Emergency response can be characterized as quick, appropriate action taken to protect people, save lives, and improve the well-being of refugees. Emergency management will prioritize life-saving measures after safe refuge is guaranteed. The areas where there are still gaps from the contingency plan, which will subsequently be converted to an operational plan, will be confirmed or identified with timely and quick evaluations of issues, requirements, and resources. It is critical to identify issues that require specialized knowledge. For most refugee situations, in addition to protection personnel, community services personnel and one or more technical specialists will be needed to manage the vital technical sectors, such as infrastructure, sanitation, water, sanitation, food, and infrastructure (United Nations High Commissioner for Refugees 2007).

4.2 Contingency Planning

Contingency planning is a forward-thinking process in an uncertain environment where scenarios and objectives are agreed upon, managerial and technical actions are defined, preparedness measures are taken to lessen the effects, and response systems are implemented to better prepare for or respond to an emergency. Building organizational capability through contingency planning lays the groundwork for operations planning and all other aspects of emergency response (Fig. 4.1).

The objective of contingency planning is to identify and assess potential emergencies and define appropriate responsibilities and actions. A team representing UNHCR and partner organizations (a Planning Group) must collaborate during contingency planning. The Planning Group must then follow up to ensure implementation. Rapid and efficient emergency response requires contingency planning. In the early stages of an emergency, significant time will be lost without proper contingency preparation. The goal is to arrange current funds and resources and identify remaining funds in the annual budgeting program (United Nations High Commissioner for Refugees 2007).

Prior to the rise of the Islamic State of Iraq and the Levant (ISIL) in Mosul, the city boasted a population of 1,377,000 residents. However, in June 2014, the city fell to ISIL, and this led to a devastating impact on its inhabitants, with an estimated 500,000 people compelled to flee their homes. Anticipating the need for a massive and intricate humanitarian response in 2016, prominent aid organizations took

Fig. 4.1 Contingency planning process

proactive measures. As early as February, they began formulating a contingency strategy, preparing for the worst-case scenario where up to 1.5 million individuals might require assistance. A planning group was formed with key stakeholders such as the CCCM cluster, civil-military coordinators, and various other organizations and partners to ensure effective coordination and collaboration. Thanks to extensive coordination efforts among the members of this planning group, they managed to garner support and generate a collective commitment to provide essential humanitarian assistance at various sites across the region.

Every disaster preparation strategy must begin with an early warning. In order to indicate the seriousness of a possible emergency, the UNHCR Headquarters in Geneva (HQ) maintains an early warning action alerts system that divides nations into four colour-coded groups. These are (United Nations High Commissioner for Refugees 2007) as follows:

- Red indicates probable crises that are anticipated to occur in the following 3 months.
- Orange indicates possible emergencies that are anticipated to emerge during the next 6 months.
- Yellow indicates possible emergencies that are anticipated to occur during the following 12 months.
- Blue indicates nations where there is still no expectation of a crisis in the upcoming year.

Field offices and HQ desks keep an eye out for indications of a potential emergency that could result in the influx of refugees. This is done by examining various sources, including local politics, political leaders, media, academics, refugees, and international and national organizations. The action alerts system changes the classification of countries in accordance with analysis conducted in the field, at HQ, and elsewhere. When a country receives a yellow or higher classification, contingency planning or current plan review should start. Planning for an emergency that does not occur is preferable to not having a plan when one is needed (UNHCR n.d.-a).

Regular meetings are necessary for the contingency planning process to monitor the implementation of the preparedness recommendations made in the plan and follow up on the first draft plan. These sessions should also assess the presumptions, indicators, and scenarios included in the plan and make any necessary adjustments to reflect how the situation has changed. A plan incorporating the following should be the result of a meeting for contingency planning (UNHCR):

- Scenario analysis
- Assumption and indicator development
- Strategic goals
- Sector goals and actions
- Resources needed for an action plan
- Recommendations for safety precautions

The present strategy should be updated at subsequent meetings after reviewing early warning indicators and making any necessary adjustments to the scenarios.

It is helpful to identify local environmental issues or concerns in advance to include them in a contingency plan. These issues or concerns may be relevant to the planned or ongoing activity. Creating such a site-specific plan can aid in preventing or at the very least minimizing, adverse environmental effects and help identify environmental risks that might affect the health of refugees.

A good plan achieves the ideal balance between thoroughness and minimal detail, considering all the pertinent aspects without being overly complex. It must be organized, simple to read, and, most importantly, simple to update. It should be clearly laid out with indications of what must be done, by whom, and by when (United Nations High Commissioner for Refugees 2007).

4.3 Temporary to Permanent Living

People's increased susceptibility to extreme physical events may be seen as being closely related to the world's ongoing refugee crisis, which is characterized by the increase in populations with extremely low living standards and the concentration of resources under the control of a minority. Emergency architecture offers two different kinds of intervention: one that is immediate and uses temporary emergency shelters during territorial rehabilitation, and a second that is permanent and involves rebuilding the homes that were lost when the disaster occurred, which also entails rebuilding the surrounding area. This first style of emergency architecture is thought of as being transient, quick, and inconclusive, intending to not last for a very long time. The shelter must be seen as a process, not as a destination, thus being occupied temporarily and should only be used in an immediate situation. When designing a space, the architect must remember that people will eventually need to find a place to call home and reintegrate into society. However, an architect's duties go beyond only constructing spaces. The architect must be sympathetic and consider the requirements of those in need of refuge while always being mindful of cultural differences, the location, and the resources available there. It is also essential to make it apparent that the shelter is something transient and not intended to be permanent.

Moreover, increasing sensitivity to catastrophic physical disasters has far-reaching social repercussions. One social effect is increased inequality. The wealth disparity widens due to a minority's control of resources and displaced masses' vulnerability. Refugees, frequently without financial or social capital, are further alienated and unable to cope with or recover from harsh physical disasters. Inequality perpetuates poverty and displacement, with the poorest suffering the most from environmental and social catastrophes.

Additionally, the lack of access to basic services and resources due to extreme events increases community conflicts. Refugee camps and host towns often clash over food, water, and shelter. Resource scarcity sometimes causes prejudice, marginalization, and social unrest between refugees and locals, isolating vulnerable groups.

Extreme physical events add to migrants' psychological and emotional suffering. Their stress and instability from natural disasters increase their risk of anxiety, sadness, and other mental health concerns. Due to their concentration on survival, refugee populations typically struggle to build supportive social networks that help them cope.

Furthermore, the increased displacement caused by extreme physical events can overwhelm host countries and regions, leading to overstretched resources and infrastructure. This can produce societal instability as governments and humanitarian groups struggle to meet the demands of expanding refugee populations, making refugees feel abandoned and neglected. Refugees may be seen as a drain on host communities' resources, causing xenophobia and social discord.

4.4 Operation Planning

Well-planned emergency response is necessary. Creating an operations plan is a significant factor in planning, especially in a crisis. An evaluation of the challenges, needs, and resources should serve as the foundation for the "Operations Plan," a vital management tool. The plan should specify actions that need to be taken by the actors accountable for the various sectors of an operation as well as program priorities, objectives, and goals. In an emergency, specific duties and the parties in charge of carrying them out must be determined, and a clear, detailed plan must be developed. Planning often gets put off at the beginning of an emergency because knowledge is lacking, and immediate demands can be fulfilled piecemeal without a strategy. This tendency should be avoided. The operations plan should ideally utilize the contingency planning procedure. Below are the tasks of operational planning (United Nations High Commissioner for Refugees 2007):

- Assess the contingency plan's information and existing plans.
- Analyse issues, protection hazards, resource demands, and significant unmet needs by employing analysis of age, gender, and diversity. Participatory assessments of the issues and risks associated with protection, needs, and availability of resources determine what needs to be done and where the priorities should be. Plans must be revised to account for new assessments and implementation progress. It is required to determine the criteria by which help should be provided by using the findings of the participatory assessments to identify major unmet requirements and compare them to established norms. It is also necessary to identify the resources that are needed and those that are already available. Resources include employees, local and international partners for implementation and operations, and material resources.
- The entire operation and strategic objectives must be specified and clarified regarding protection. All further goals and initiatives should be in line with these broad aims. The intended consequence should always be considered when setting objectives. The period within which goals are to be attained should also be mentioned, and goals should be explicit, quantifiable, achievable, and practical.
- Clarifying the fundamental restrictions, planning presumptions, and guiding principles of the emergency operation will also be necessary. The government's and operational partner's role, duties, and policies should all be explained in precise detail. Standard or established procedures should also be outlined, including monitoring and coordination systems. Similarly, if the plan includes objectives, outputs, and actions for different sectors separately, it should also specify standards in those sectors and any necessary instructions.
- Determining the courses of action to reach overall objectives (implementing arrangements) is significant for operational planning. Planners should consider many possibilities to achieve goals, their benefits and drawbacks, and which ones are flexible, the most efficient, and the most successful. In a changing situation, it is crucial to select an alternative for executing agreements that maintains flexibility.
- The planners should decide on sector-level goals and strategies for achieving those goals. They need to make sure that every sector defines precisely how the various needs of women, girls, boys, and men will be met and prioritize targeted action to empower women and other discriminated groups.
- It is essential to clearly define roles within the humanitarian organization and amongst the various participants in the operation.
- Mechanisms for coordination amongst the various operation actors should be established. Additionally, coordination at various geographic levels (such as the location and the regional or capital city) must be ensured. Various sector-specific coordinating systems could be required for a large operation.
- Managing a refugee emergency must include ongoing monitoring (measuring performance indicators) in collaboration with the community, reporting, and evaluation from the beginning to ensure that the objectives remain effective as circumstances change.

- The planners and management should document and distribute the plan, track the project's development, identify problems, act to fix them, and then update and amend the plan (United Nations High Commissioner for Refugees 2007).

4.5 Computational and Parametric Design for Rapid Delivery

The use of computational design is expanding in the building, engineering, and architecture sectors. Recent developments in computer architecture in scholarly study and technological innovation are reflected in "Parametric Modeling". It enables the designers to manage the creation of 3D objects that can be seen from a larger logical computing script or scenario (Daniel et al. 2011). They incorporate mathematical formulas, restrictions, and control functions to generate a geometric model from input data. This method's distinctive feature is creating models with various simple shapes rather than static geometry (Fernando et al. 2012). In the case study of the Hex shelter (Sect. 4.6), parametric design techniques were utilized to generate scalable shelters that are not only rapidly deployed and erected by unskilled builders but are also dignified with typical home amenities.

In times of crisis and calamity, the demand for shelter becomes more pressing, and the design must adhere to several requirements (technical and sociological). As a result, the design must also consider contextual and climatic factors, such as site circumstances, resident culture, traditions, and climate, in addition to physical factors like gross size, height, and structure. Using various factors, generation tools can be the best way to address this growing need (Daher et al. 2015).

It is possible to relate parametric digital design approaches to the constraints and requirements of designing emergency camps and shelters. The constraints established at the level of both camps, and shelters should be reflected in the variables and parameters (Daher et al. 2015).

4.5.1 Shelter Design Generation

The process of designing a parametric shelter is linear. Figure 4.2 shows the steps for the generative shelter design process to comprehend the logical sequence in which information relating to these various elements is defined and subsequently inserted in parametric modelling approaches.

The first step is determining how many people will reside in the shelter. This parameter makes it possible to calculate the needed surface shelter using the average area per person. The minimum area required by the UNHCR Camp Planning Standard is 3.5 sq. m. (37.67 sq. ft.) per person. In cold climates and urban areas, more than 3.5 sq. m. (37.67 sq. ft.) may be required. Moreover, a minimum ceiling height of 2 m (6.56 ft.) at the highest point will determine the interior size of the

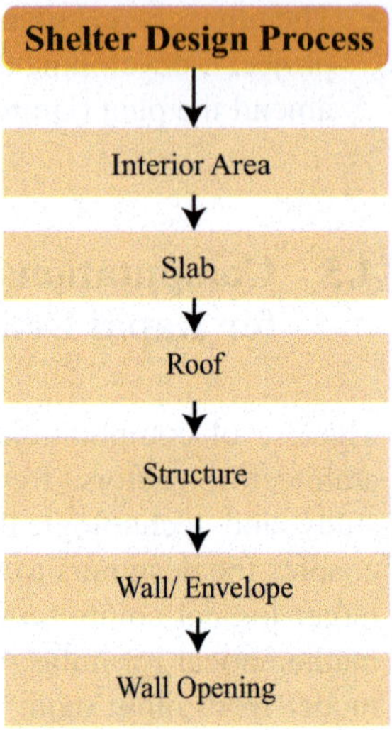

Fig. 4.2 Steps showing the shelter design process

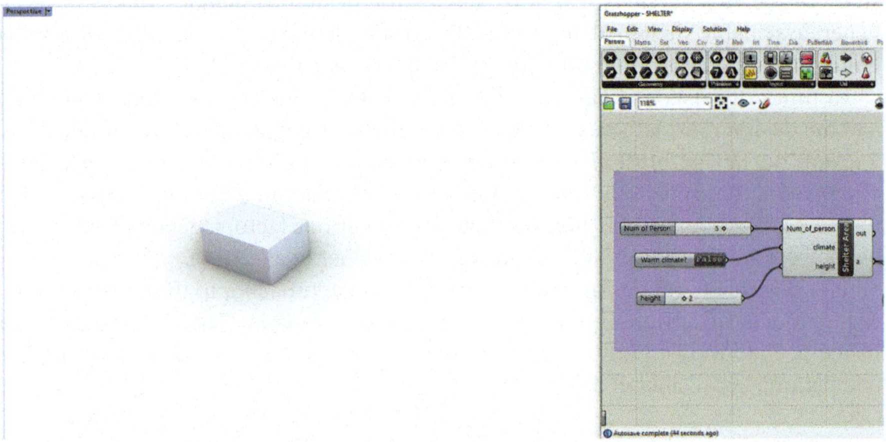

Fig. 4.3 Shelter design generation using generative design tool (Grasshopper)

shelter required by each family. Using these parameters, the design tool will generate the shelter volume for each family (Fig. 4.3).

4.5.2 Refugee Camp Master Plan Generation

Emergency camp design is a more complex and non-linear process. New concepts, requirements, and ideas will constantly be introduced, and planners should be prepared to adjust accordingly.

The design of the camp should be linked to the types of solutions envisaged for sheltering, which must conform to standards. First, refugee camp planning standards by UNHCR should be listed and followed as the camp planning parameters. Below are the steps that should be followed for camp design:

Step 1. Generating Camp Settlement Size

According to UNHCR camp planning standards, the camp settlement size is 45 sq. m. (484 sq. ft.) per person (including kitchen and vegetable garden). The refugee camp area needed for the incoming number of refugees can be determined by multiplying the number of people by 45 sq. m. (484 sq. ft.) per person (Fig. 4.4).

Step 2. Diving Camp Settlement into Smaller Module

The camp settlement is then divided into smaller modules (i.e. sector, block, and community) according to the UNHCR camp planning standards (Table 4.2). The camp settlement size will be divided into smaller modules depending on the number of refugees (Fig. 4.5).

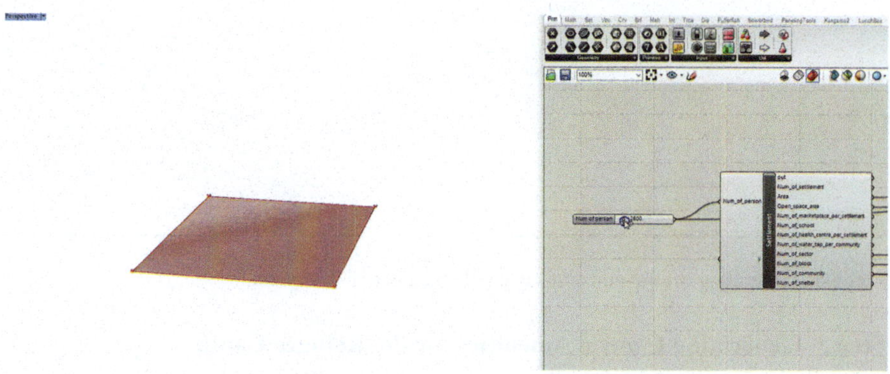

Fig. 4.4 Generating camp settlement size using the generative design tool

Table 4.2 Modular planning units

Module	Structure	Approximate number
Family	1 × family	4–6 persons
Community	16 × families	80 persons
Block	16 × communities	1250 persons
Sector	4 × blocks	5000 persons
Settlement	4 × blocks	20,000 persons

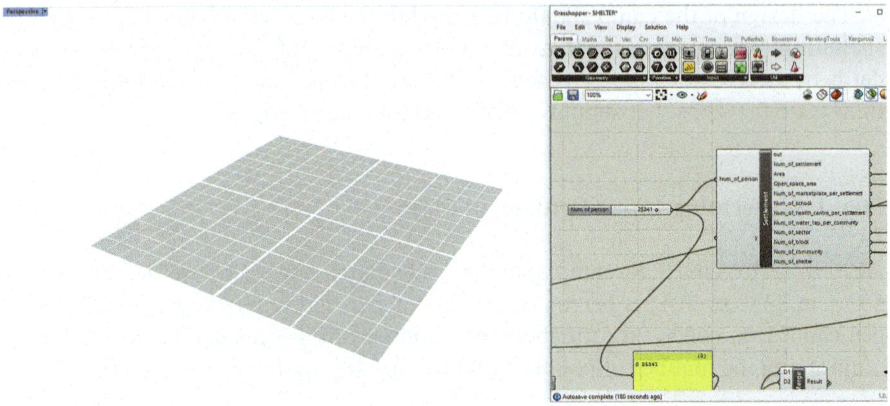

Fig. 4.5 Dividing the camp settlement into smaller modules using the generative design tool

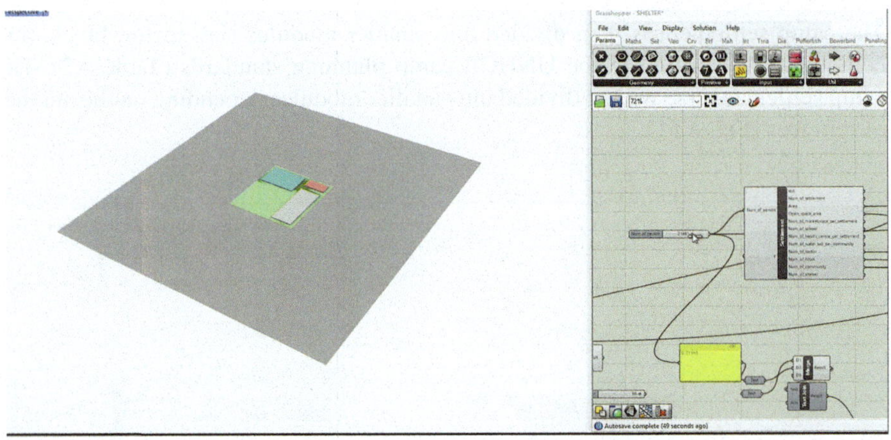

Fig. 4.6 Generating central amenities using the generative design tool

Step 3. Generating Central Amenities for the Refugee Camp

The generative design tool will generate the amenities needed for the refugee camp based on the number of refugees. UNHCR camp planning standards have provided guidelines for this number, which can be used to generate the parametric refugee camp design (Fig. 4.6).

Step 4: Generating the Road Network and Open Spaces for Camp Communities

After generating the central amenities and camp communities, roads need to be generated hierarchically by primary, secondary, and tertiary. Then, the community's open spaces and shelters are distributed. Finally, the sub-community amenities are

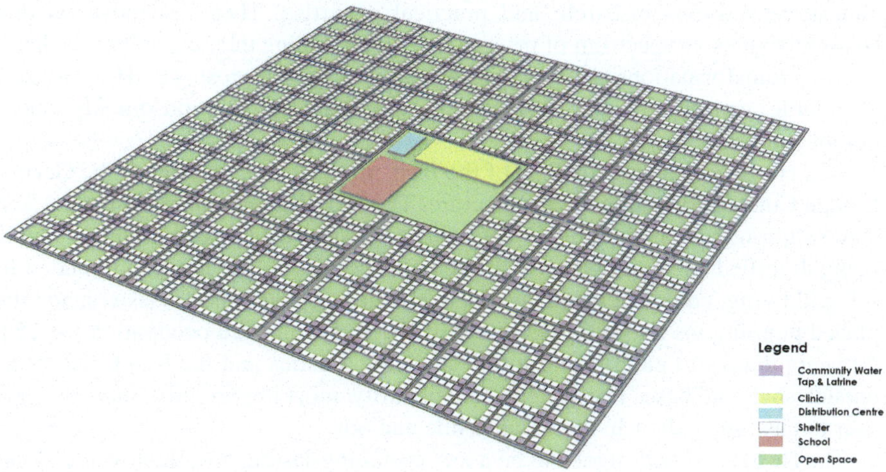

Fig. 4.7 Final refugee camp design generated from computational design tool

located within the community design and should serve the needed facilities (i.e. water taps and latrines) for the people living within the communities.

Parameters for shelters and camps are utilized to evaluate this research's hypotheses are incorporated in a 3D modelling program that enables the creation of a prototype while considering parameters and constraints. This study uses Grasshopper and Rhino 3D because these software programs enable it to develop and manipulate algorithms based on visual objects and define the constraints and parameters. At the camp level, the scenario experimented with in this case considers flat land. However, computational design tools can be used for hilly regions.

The prototype created using the tools should be tested and evaluated by professionals in the fields of architecture and urban design and the humanitarian sector. The review should be conducted through several experiments depicting various scenarios quickly formed by adjusting the settings to test the design process's capacity to respond to scenarios that are similar to actual cases (Fig. 4.7).

4.6 Case Study: Low-cost Generative Hexagonal Shelters Cluster for Refugees

Architects for Society set out in 2015 to build housing solutions that aid in alleviating and overcoming many of the issues connected with the present temporary refugee shelters in light of the prevalence of refugee displacement around the world. Architects for Society (AFS) is a non-profit design organization dedicated to a noble mission: empowering underserved communities through visionary architecture and innovative design solutions. Hailing from diverse corners of the world—the USA, Canada, Europe, the Middle East, and India—these seasoned architects unite under the AFS banner with a shared commitment to design education, rigorous research, and practical expertise. Their collective wisdom spans an extensive spectrum of project types, reinforcing their capacity to usher in positive transformations within communities that need it most. The Hex House is affordable, can be delivered in pieces for the customer to build quickly, and is durable and respectable.

Hex House was awarded "Best of the Best" in the Iconic Awards by the German Design Council for visionary architecture. The Hex House is intended to be a scalable solution that can be quickly deployed and constructed by non-builders while being dignified and equipped with basic home comforts. The home is intended for off-grid living, emergency housing during natural or man-made disasters, and displaced populations like refugees. It can withstand prolonged occupation for 15 to 20 years thanks to passive and low-tech characteristics and the toughness of the outside shell and base. The system's adaptability, most importantly, allows the end user to customize their living areas inside and out.

The Hex House is designed to be a low-cost, long-lasting, fast-deployable, dignified home that can be transported in parts and assembled by the end users. The fundamental construction elements consist of structural insulated metal panels for the walls, floor, and roof, which may be modified, and galvanized tube steel for the base.

The hexagonal shape's inherent structural stability and the construction of inflexible structural insulated panels forbid the use of additional structural support. The wall and roof panels have inherent tongue and groove connections that bind them to create a robust structural shell. They are designed to be self-supporting.

Hex housing clusters were designed using generative design software. Several cluster configurations may be created using the 40 sqm hexagonal layout and around common outside gardens or green spaces. The generative cluster using computational parametrics made the design of the clusters unique and variable. These clusters can be arranged to enable accessible roadways and pedestrian pathways, whether linear or radial. For improved thermal performance, units can be linked together and shared walls, or placed next to one another in suitable patterns. Additionally, they may be merged to create bigger units.

A two-bedroom house with all the conveniences designed for small families is 40 sq. m. (431 sq. ft.). To allow for plan modification and natural ventilation, inside walls are not fastened to the ceiling (Fig. 4.8).

The Hex House's straightforward building process makes it easy for residents to complete assembly with only a few simple tools and instructions. Each component is made to be flattened out and delivered in a standard transport trailer. Three units can readily fit into the normal trailer's capacity.

Rainwater is directed through an integrated gutter system and discharged via a downspout system. It is filtered and manually pumped back into the home by point-of-use hand pumps for flushing and washing in a sub-grade storage tank. Additionally, there is a below-grade refillable potable water storage tank that is also manually pumped. Due to the earth's inherent heat insulation and the house itself, both tanks are protected (Fig. 4.9).

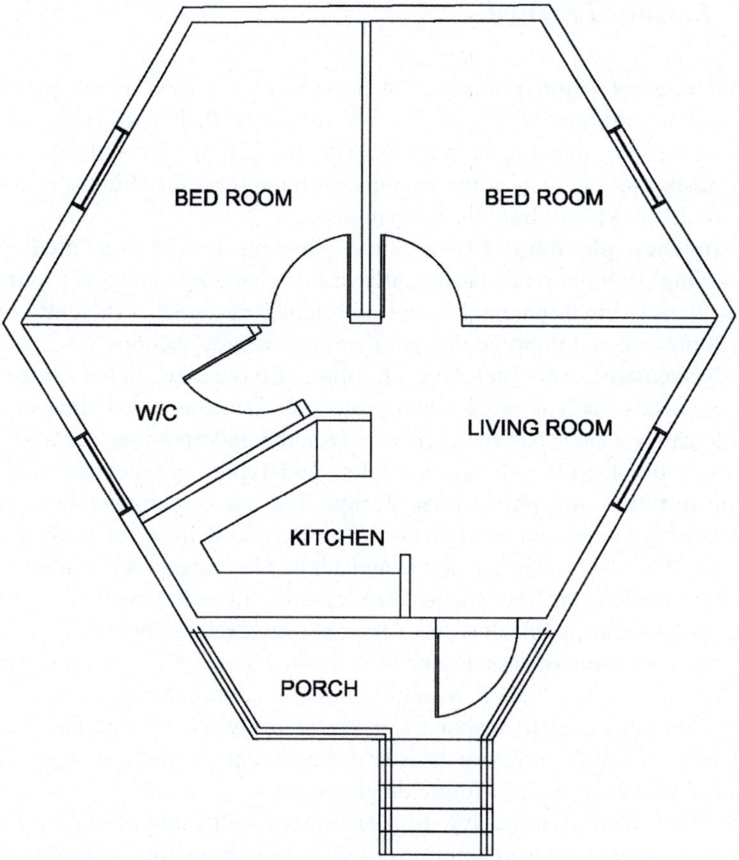

Fig. 4.8 Floor plan of Hex shelter

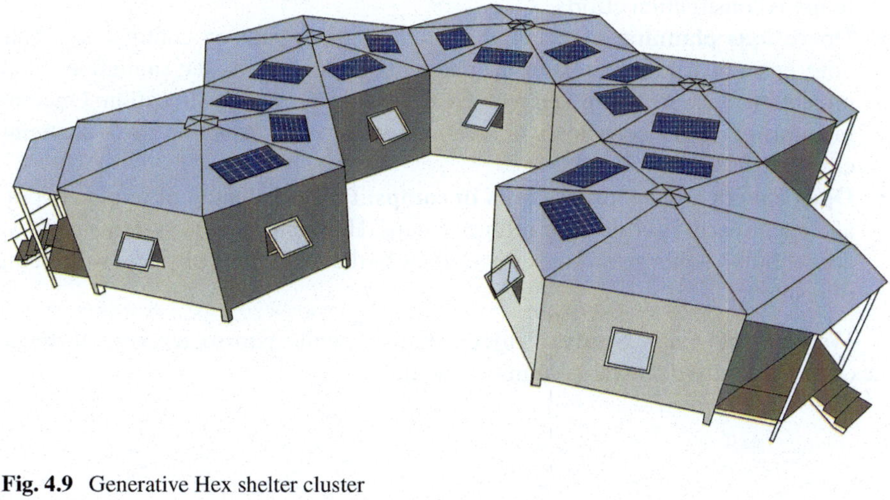

Fig. 4.9 Generative Hex shelter cluster

4.6.1 Lessons Learned

- **Importance of rapid planning**: In humanitarian crises, timely planning and coordination are crucial for providing immediate shelter to displaced populations. Effective planning helps anticipate risks, minimize delays, and ensure swift responses, as seen in the example of building refugee housing for 17,500 households in Mosul, Iraq, within 6 months.
- **Contingency planning**: Contingency planning is vital for predicting and addressing potential risks, particularly in fast-changing emergency situations. It helps identify environmental issues and deploy resources efficiently to reduce execution time and improve the quality of emergency shelters.
- **Gender-sensitive and inclusive planning**: Emergency shelters must address specific needs, such as privacy and protection for women and children, to safeguard against gender-based violence and trauma. Incorporating separate sleeping and sanitation facilities ensures the safety and dignity of vulnerable groups.
- **Computational and parametric design**: The use of computational and parametric design approaches revolutionizes the rapid delivery of refugee housing. These technologies enable scalable and adaptable designs, accommodating various requirements such as shelter size, community infrastructure, and environmental conditions, while ensuring efficient use of resources.
- **Emergency preparedness**: Preparedness is key to effective emergency response. Organizations like UNHCR maintain globally coordinated resources and regularly train personnel to respond quickly to disasters. Preparedness measures, such as Cambodia's proactive disaster management, demonstrate how prior planning reduces the impact of future emergencies.
- **Transition from temporary to permanent solutions**: Temporary shelters should be seen as interim solutions, with the ultimate goal being to reintegrate displaced people into permanent, safe housing. Planning for this transition ensures that temporary shelters provide dignity while allowing flexibility for future reconstruction efforts.
- **Operations planning**: A detailed operations plan, built on contingency planning, ensures that all sectors of the emergency response (water, sanitation, food, protection, etc.) are addressed with clear goals and resource allocation. Ongoing monitoring and coordination are essential for adapting to changing circumstances.
- **Open spaces and road networks in camps**: Proper planning of road networks and open spaces is critical for refugee camps. Hierarchical road systems improve accessibility, while green spaces contribute to the mental and physical well-being of residents.

Weblink of Case Study Project: https://architectsforsociety.org/our-work/hex-house-rapidly-deployable-dignified-home/

4.7 Conclusion

Delivering living spaces for refugees during the emergency period can take a significant amount of time. The major hurdles to be overcome are rapid planning, construction, and organizing construction costs. Nowadays, using computational design tools, designers can generate the refugee camp design and produce a rapid design for implementation. Planning ahead and following contingency planning strategies can solve many unseen problems that can delay the construction and implementation of shelters and refugee camps. Moreover, operational planning is crucial for developing implementation strategies.

Discussion Questions

- What are the key considerations for providing immediate shelters following a conflict or environmental disaster?
- How do flat-pack emergency shelters contribute to rapid project delivery and installation?
- What role does contingency planning play in the rapid and effective installation of emergency shelters?
- How can identifying environmental issues during contingency planning help in the design of shelters and camps?
- What are the benefits of using computational and parametric design in the planning and delivery of emergency shelters?
- What are the challenges and strategies involved in transitioning from emergency shelters to long-term living solutions for displaced people?

References

Daher E, Kubicki S, Halin G (2015) A parametric process for shelters and refugees' camps design. eCAADe 2. https://doi.org/10.52842/conf.ecaade.2015.2.541

Daniel D, Jane B, Mark B (2011) Untangling parametric schemata: enhancing collaboration through modular programming. In: Leclercq P (ed) CAAD futures 2011: designing together. University of Liege, Liege

Fernando R, Drogemuller R, Burden A (2012) Parametric and generative methods with building information modelling: connecting BIM with explorative design modelling. Beyond Codes and Pixels—Proceedings of the 17th International Conference on Computer-Aided Architectural Design Research in Asia, CAADRIA 2012, p. 537–546. https://doi.org/10.52842/CONF.CAADRIA.2012.537

Global Shelter Cluster (2021) Shelter projects essentials-learning from programming in humanitarian crises. chrome-extension://efaidnbmnnnibpcajpcglclefindmkaj/https://shelterprojects.org/essentials/shelter-projects-essentials-2021-print.pdf. Accessed 30 May 2024

Humanitarian Response (2021) Emergency preparedness and response plan. chrome-extension://efaidnbmnnnibpcajpcglclefindmkaj/https://www.un.org/unispal/wp-content/uploads/2021/12/OCHAHUMRESPLAN2022_161221.pdf. Accessed 16 May 2024

International Organization for Migration (IOM) (2023) Shelter projects. https://www.google.com/search?q=Humanitarian+response+(2021)+Emergency+Preparedness+and+Response+Plan&rlz=1C1CHBF_enCA817CA817&oq=Humanitarian+response+(2021)+Emergency+Preparedness+and+Response+Plan&gs_lcrp=EgZjaHJvbWUyBggAEEUYOdIBCDIzNDRqMGo3qAIIsAIB&sourceid=chrome&ie=UTF-8. Accessed 16 May 2024

UNHCR (n.d.-a) Preparedness package for refugee emergencies (PPRE)—UNHCR|Emergency Handbook. https://emergency.unhcr.org/entry/34912/preparedness-package-for-refugee-emergencies-ppre. Accessed 16 Aug 2022

UNHCR (n.d.-b) PPRE contingency planning (refugee situations)—UNHCR|Emergency Handbook. https://emergency.unhcr.org/entry/35282/ppre-contingency-planning-refugee-situations#1,1660794853153. Accessed 16 Aug 2022

United Nations High Commissioner for Refugees (2007) Handbook for emergencies. United Nations High Commissioner for Refugees, Geneva

Chapter 5
Infrastructure and Utilities

Keywords Refugee camp design · Infrastructure and utilities for camp planning · Refugee camp drainage systems · Water supply in refugee camp · Sanitary sewers in refugee camp · Solid waste management in refugee camp · Wastewater treatment in refugee camp · Refugee community facilities · Fire safety in refugee camp · Rohingya refugee camp · Emergency shelter planning

In the design of a refugee camp, careful attention must be given to adequate infrastructure and utility components such as drainage, water, and sanitary systems since they affect inhabitants' lives. Besides housing infrastructure, communal facilities such as community gathering spaces, schools, clinics, and administrative centres are necessary as long-term needs in a refugee camp. Utilities should be given the highest priority and be considered from the beginning of the site planning phase. Every refugee camp has needs and conditions for utilities and infrastructure that must be met to ensure healthy living.

This chapter focuses on the essential infrastructure and utilities necessary for providing a safe and sustainable environment within refugee camps during emergencies. The initial establishment of shelters is followed by the construction of physical and social infrastructure, while challenges arise due to unpredictable administrative and municipal needs. The difference between conventional and refugee camp development lies in the semi-permanent nature of infrastructure and utilities. The drainage system plays a pivotal role, with proper planning required to ensure efficient water management, minimize erosion, and prevent flooding. Flood-prone areas must be avoided, as improper drainage can damage shelters and access routes. Sanitary sewers, solid waste management, wastewater treatment, and pest control are vital aspects of maintaining hygiene and preventing diseases within the camp. Clean water is a basic human right ensured through comprehensive water supply systems. The chapter also underscores the significance of fire breaks, community facilities like schools, places of worship, and marketplaces in promoting community well-being and engagement. A Rohingya Refugee Camp case study in Bangladesh illustrates how a whole settlement approach involving careful site

A. Friedman, B. Chaki, *Fundamentals of Planning and Designing Sustainable Post-Disaster Shelters*, https://doi.org/10.1007/978-3-031-83317-5_5

planning, development, and active community participation can effectively address infrastructure and utility needs while minimizing environmental impact.

5.1 Local and Storm Drainage System

During an emergency, usually, shelters are built first; then, physical and social infrastructure and utilities are constructed. It may be difficult to construct all of the administrative and municipal services needed at the onset of an emergency. Moreover, predicting the eventual number of refugees and the infrastructure required may be difficult. Therefore, empty spaces should be allocated for these services' inclusion or future growth. The difference between conventional development and refugee camp development is that the infrastructure and utilities provided in the camp are semi-permanent types. The drainage system for a campsite should be designed in parallel with the site plan. Integrating site infrastructures, including drainage and water, with site planning as a "whole settlement" is significant. The Rohingya refugee camp case study in Sect. 5.6 is a great example of how the camp planners integrated all infrastructure components cohesively to meet the community's overall needs.

Contours identify the ideal infrastructure locations for steeper land. Major roads should follow drainage systems to guarantee that slopes are respected, that floods do not occur, and that infrastructure expenditures are minimized. As long as the gaps between fire breaks do not become too big, fire breaks should be integrated into the drainage system. Using fire breaks to divert surface water run off can help to decrease erosion and gullies on paths and within shelter blocks .

As noted in Chap. 2, flood-prone areas should be avoided during the site selection stage. If a camp is inadequately drained and prone to floods, shelters can be damaged or destroyed, and routes to amenities can get clogged. Any strategy to reduce flooding in a camp should prioritize measures for the most severely impacted areas (Global Shelter Clusters 2018).

A *storm drainage system* collects and directs rainfall away from shelters and roads. Each shelter transfers water collected around its perimeter to the drainage system. When proper soil grading is done, rainwater from the site will be directed into the road to *catch basins* that will further move the rainfall through pipes into a large body of water, such as a river. The catch basins are spaced apart near the curb (Ajibade et al. 2019).

When exploring cost reductions for storm drainage systems, an underlying principle of all strategies must be the reduction or even elimination of costly pipe networks. Once built, they have to be equipped with manholes and constantly maintained. Their elimination can be achieved by properly grading to natural features such as landscaped areas (Ajibade et al. 2019).

In addition, two tier systems can be conceived: one that can handle rain from a normal storm that happens every year or two and serves a residential area, and a second that can handle storms occurring every 5–10 years and will be constructed around major traffic arteries and commercial areas. Another economic approach to handling stormwater is the creation of retaining ponds, which are far less expensive

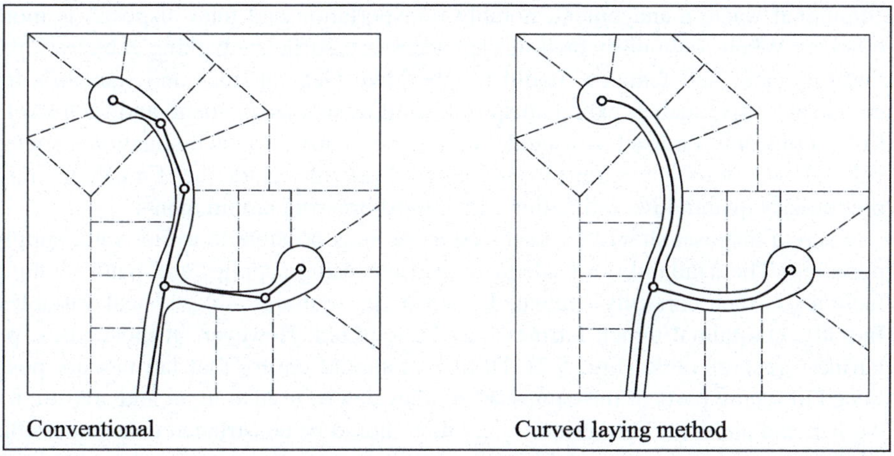

Fig. 5.1 Conventional and curved laying method for manholes

than a drain system. Having ponds or lakes at the centre or the periphery of development can provide a recreational facility for a community. Another economic strategy is the building of a continuous retention ditch at the side of the road. With proper grading, the ditch—which may also have a landscaped berm—will absorb the water or lead it to a retaining pond in the case of an overflow (Ajibade et al. 2017).

When conventional storm drain systems with pipes and manholes are planned, less costly materials and construction techniques can be used. With the invention and proliferation of plastics, the traditional metal and reinforced concrete materials are being phased out. Polyvinyl chloride (PVC) is also corrosion-resistant and can be installed in most "minor" storm systems. The spacing and construction techniques of manholes can also be questioned. Some municipalities still require old standards of spacing manholes 60 m (200 ft.) apart. Improved construction methods permit extending them to 180–240 m (600–800 ft.). Some manholes can be eliminated by constructing curved sections as it requires fewer connection points and changes in direction points (Fig. 5.1). The use of a single-unit precast concrete manhole can also yield savings (Balbastre-Soldevila et al. 2019).

5.2 Sanitary Sewers

Examples of sanitation include water quality control, appropriate disposal of human excreta, wastewater, rubbish, dead bodies, insect and rodent control, safe food handling techniques, and good site drainage. Public and environmental health improvements require such interventions, and appropriate personal, home, and communal hygiene habits. All of these activities and the provision of health care are intricately linked and should be evaluated as a whole (Kochary 2020).

Communal facilities, particularly latrines, are difficult to keep clean and maintain due to the extensive number of uses by the refugee community. However, communal waste management, notably transportation and final disposal, is more efficient. When community facilities are necessary, sufficient bathing cubicles separate for males and females should be provided. General rules and standards for certain activities such as excreta disposal, solid waste management, and vector control should only be used as a guideline and be adapted to each situation's social, cultural, and physical circumstances. Table 5.1 provides guidelines for calculating a preliminary quantitative estimation of the most pressing requirements.

Excreta disposal should be localized as soon as possible to avoid water supply pollution. The availability of adequate anal cleansing supplies and hand washing facilities is then carefully evaluated, considering cultural and physical variables. Initially, communal trench latrines may be required. However, in most cases, pit latrines are preferable (Fig. 5.2). Designers should ensure that latrines are positioned in secure parts of the camps where they can be used at night and are safe for women and children. Furthermore, facilities should be constructed to be accessible to all individuals, including children, the elderly, pregnant women, and persons with disabilities or illnesses.

5.2.1 *Solid Wastes*

Insect and rodent-borne illnesses are increased by improper waste disposal. Hence, a garbage storage, collection, and disposal system must be developed. Garbage disposal places should be marked, and access to them should be limited. Dust and/or smoke in large quantities can harm one's health. Preventing the destruction of vegetation is the greatest way to avoid dust (Harvey 2004).

At the individual home level, metal drums can be utilized as garbage cans. A 757-L (200-gallon) drum chopped in half is commonly utilized. If at all feasible, bins should have lids and drainage holes at the bottom. It has been proven that a ratio of one container (100-L (26 gallon) capacity) for ten families is effective. The containers should be positioned throughout the site such that no residence is more than 15 m (49 ft. 2 in.) distant from one (United Nations High Commissioner for Refugees 2015).

Table 5.1 Minimum requirement for sanitary facilities (United Nations High Commissioner for Refugees. 2015)

	First option	**Second option**	**Third option**
Excreta disposal	1 latrine per family (or 1 latrine per two families)	1 cubicle per 20 persons	1 cubicle per 100 persons or a defecation field
	Storage	**Transport**	**Final disposal**
Refuse/garbage	1 bin of 100 L per 10 Families or 50 persons	1 wheelbarrow per 500 persons and 1 tipper per 5000 persons	1 landfill (50 m^2 and 1.2 m deep) per 500 persons and 1 incinerator and 1 deep pit for each clinic

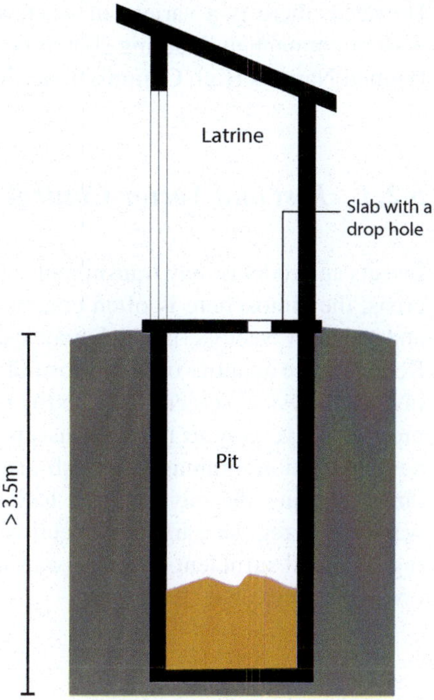

Fig. 5.2 Pit latrine section

5.2.2 *Wastewater*

Wastewater must be managed, and drainage should be provided as quickly as practicable. Wastewater pollution of the surrounding ecosystem must be avoided at all costs. This matter of cleanliness should always be taken into account from the start. Drainage keeps water from becoming stagnant at distribution points by draining rainwater and residential wastewater from diverse sources (toilets, showers, and kitchens). Eliminating ponds is another method that can help control vectors. At the following locations, quality drainage should be a top focus (United Nations High Commissioner for Refugees 2015):

- Water points (standpipes, taps, and handpumps).
- Wastewater from sanitary facilities should be utilized to irrigate vegetable gardens and fruit trees or drained into absorption trenches or soak-away pits.
- While most household members can protect their shelters from runoff waters by using perimeter drains, it is still necessary to guarantee that such water is collected and disposed of through main drains.
- Water used to clean slaughter slabs at marketplaces and slaughterhouses.

For example, wastewater from sewers collecting effluent from pour-flush toilets needs to be treated in some cases. Certain treatment packages are available on the market, but they are generally costly, complicated, and difficult to run and maintain.

However, there is a variety of wastewater treatment technology. Professionals in waste treatment engineering should be contacted while choosing the best technique (United Nations High Commissioner for Refugees 2015).

5.2.3 Pest and Vector Control

Insects and rodents may transmit and spread diseases and degrade food. In a refugee crisis, the atmosphere is often conducive to the spread of disease-carrying insects and rodents ("vectors"), which may also damage or ruin enormous amounts of food. Reducing the quantity of flies, mosquitoes, and rodents in an emergency is challenging, and physical screens may be the best immediate response. In the long run, the most effective way to manage insects and rodents is to enhance personal cleanliness, sanitation, drainage, rubbish disposal, and food storage and handling procedures, making the environment less conducive to vectors. Eliminating stagnant wastewater, regular garbage collection, and controlled excreta disposal and providing soap and sufficient water for washing are all examples of practical approaches (WHO-Geneva et al. 1988).

5.2.4 Dead Body Disposal

From the beginning of a refugee crisis, proper plans for the disposal of the deceased are essential. The death rate may be higher than under "normal" circumstances. The authorities should be called immediately to guarantee conformity with national processes and, if needed, for help. Unless the cause of death was typhus or plague, in which case they may be infested with infectious lice, fleas, or cholera, dead bodies pose no health concern.

Funerals for cholera victims should be held swiftly and close to the site of death. Efforts should be taken to limit the number of people attending funeral gatherings for people who died from any of these three illnesses. Caution should also be extended to ceremonial washing of the deceased, through intense health education or regulation, if needed. Health reasons do not justify cremation, which is sometimes difficult to achieve due to a lack of fuel. The traditional manner of disposal should be used if possible, and traditions and rituals should be allowed. Material requirements, such as for shrouds, should be satisfied. The amount of space required for burial must be considered at the site planning stage, especially in congested areas (Atlas 1984).

5.3 Water Supply

Water is a basic human right since it is necessary for life, health, and dignity. Water is frequently unavailable in sufficient quantity and quality in emergencies, posing a serious health risk. As a result, water is one of the most important factors to consider when choosing a location for a refugee camp (Cairncross and Feachem 1983).

For personal hygiene and domestic requirements, at least 20 L (5.3 gallons) per person daily is required; 7 L (1.8 gallons) of water per day is the absolute requirement for survival. Water is also required for community building needs. The following is the minimum water required for communal needs (United Nations High Commissioner for Refugees 2015):

- **Health centres:** 40–60 L (10.5–15.8 gallons) per patient per day.
- **Feeding centres:** 20–30 L (5.3–8 gallons) per patient per day.
- **Schools:** 3 L (0.8 gallons) per pupil/ per day.
- **Mosque:** 2–5 L (0.5–1.3 gallons) per person per day.
- **Hand washing at communal latrines and Offices:** 1–2 L (0.1–0.5 gallons) per user per day for hand washing, and 2–8 L (0.5–2.1 gallons) per cubicle per day for cleaning.

A water supply system combines structures to extract, purify, store, distribute, and facilitate wastewater disposal. It is essential to guarantee that the system components are compatible with one another and appropriate in terms of supply and demand, and they can be maintained using locally available resources at the lowest feasible overall capital, operation, and maintenance costs.

An overall plan for the longer-term water supply system should be prepared as soon as practicable. Some aspects of the plan can be difficult to implement because of a lack of fundamental data or difficulties getting planning or design tools such as cartographical and hydrological data. Therefore, the following procedures should be maintained (UNHCR 1992).

First, probable water sources within a reasonable distance of the intended campsite should be searched. Then preliminary surveys should be conducted to determine the amount and quality of water. Following that, the outline of the construction and operating implementation plans should be prepared. The construction and operation implementation plans need to be detailed. Then, a preliminary design concept should be produced.

The organization should conduct an in-depth survey to fine-tune all features and specifics of the chosen design. Refugees' participation in the initiative is mandatory to implement the project. The implementation phase includes technical oversight, upkeep, monitoring, and reporting. Finally, an operation and maintenance plan that includes the formation of a committee composed of refugees and representatives from key aid sectors is essential for ensuring a successful water supply system for the camp.

Surface water such as streams, rivers, lakes, groundwater underground or surfacing as springs, and rainwater are the three primary natural kinds of fresh water.

Existing municipal and private water supply systems in the neighbourhood of the refugees, such as those belonging to industrial or agricultural operations, may fulfill part or all of the demand during the emergency phase and should be utilized first before pursuing alternative options. It is feasible that the yield and quality of such systems might be significantly improved. If there are no fresh water sources accessible in a given location, the refugees must be relocated as soon as possible.

A proper water distribution system should guarantee that all camp beneficiaries receive the same amount of water. A distribution system should have enough outlets to ensure users do not have to wait extended periods for service. At least one tap should be available for every 80–100 refugees, with no more than 200 people per hand pump or well with one rope and bucket. The equitable distribution of limited water is a critical topic (Sayato 1989).

Wherever practical, standpipes with push taps should be utilized as outlets. Typically, several tap standpipes are built, each installation having between two and six distinct taps. Taps are usually quite fragile; thus, spares must be kept on hand.

5.4 Fire Breaks

Generally, a 30 m. (98 ft. 5 in.)-wide firebreak (area free of buildings) is recommended for every 300 m. (984 ft. 3 in.) of built-up area (United Nations High Commissioner for Refugees 2015). Firebreaks should be placed between blocks in modular camps. This space would be perfect for vegetable gardening or recreation. If space permits, the distance between individual buildings should be sufficient to avoid collapsing and burning structures from colliding with neighbouring structures. As a result, the distance between structures should be at least twice the overall height of any structure. If the building materials are particularly flammable (such as straw and thatch), the distance should be extended to 3–4 times the entire height. Any prevailing wind's direction should also be considered (United Nations High Commissioner for Refugees 2015).

5.5 Community Facilities and Social Needs

Designing refugee housing after disasters, wars, or persecution requires careful consideration of various social aspects to ensure the well-being and integration of displaced individuals. One critical aspect is cultural sensitivity. Refugees come from diverse backgrounds, and their housing must respect their cultural, religious, and social practices. This includes providing spaces for communal activities, religious practices, and ensuring privacy, especially for women and children (Langar 2020).

Community integration is another vital factor. Housing should not isolate refugees but rather facilitate their integration into the local community. This can be achieved by designing shared spaces such as community centres, schools, and recreational areas where refugees and local residents can interact and build relationships (Ghisleni 2021).

Psychological support is crucial, as many refugees have experienced trauma. Infrastructure should include access to mental health services and safe spaces where individuals can seek counseling and support (Ghisleni 2021).

Education and employment opportunities are essential for the long-term stability of refugees. Housing designs should incorporate facilities for vocational training, language classes, and other educational programs to help refugees gain skills and find employment.

Safety and security are paramount. Refugee housing must be designed to protect residents from potential threats and ensure a safe living environment. This includes adequate lighting, secured entry points, and emergency response systems (Aburamadan et al. 2020).

Lastly, sustainability and adaptability are important. Housing should be built using sustainable materials and designed to adapt to changing needs over time. This ensures that the infrastructure remains functional and relevant as the refugee population evolves (Langar 2020). By addressing these social aspects, refugee housing can provide not just shelter, but a foundation for rebuilding lives and fostering community resilience. The master planning of the campsite should organize the facilities considering the long-term needs of the refugees. It is recommended that the plots be distinguished, including areas reserved for utilities. Additionally, infrastructure such as health centres, hospitals, schools, markets, distribution centres, and refuse drums must also be included in the master plan. During the emergency, vital outside assistance in the health sector is always required. Then in the transitional period, the infrastructure needed for long-term needs should be incorporated. The number of infrastructures needed should be determined according to the number of persons and families.

The following are some community facilities requirements suggested by UNHCR that should be followed while laying out the camp (The UN Refugee Agency 2019).

- 1 latrine per 1 family (6–10 persons)
- 1 water tap per 1 community (80–100 persons)
- 1 health centre per 1 camp (of 20,000 persons)
- 1 hospital per up to 200,000 persons
- 1 school per 1 sector (5000 persons)
- 4 commodity distribution sites per 1 camp module (20,000 persons)
- 1 market per 1 camp module (20,000 persons)
- 2 refuse drums per 1 community (80–100 persons)

5.5.1 Schools

The well-being of the whole refugee population and the social and psychological well-being of children and young people, depend on establishing an education system. Setting up leisure activities, educational systems, and other non-formal activities with the community will help to provide structure and a feeling of normalcy to displaced and traumatized people. Refugees are uprooted from not only their homes and families but also their communities. If the new community is partly based around familiar institutions such as schools, educational activities might be the group's first focal point and can generate a feeling of routine (Tyrer and Fazel 2014).

The 1951 Convention on Refugee Status and the 1989 Convention on the Rights of the Child both support the right to education for individuals of concern. Measures to encourage universal primary education, a safe school environment, and other non-formal educational activities would safeguard children and adolescents from forced labour, physical and sexual abuse, military recruiting, and other types of violence in an emergency (UNHCR 2021).

Semi-permanent structures made of semi-durable materials (i.e. bamboo and wood) can be built in safe areas to allow schools to open as soon as feasible. All refugees, including children and adolescents, should be included in determining the site of schools, latrines, and recreational places, keeping security concerns in mind. For youngsters, schools should be within walking distance. Separate latrines should be provided for females and males. The community should be organized into committees reflecting the interests of teachers, parents, and children to help develop and maintain safe school facilities. Smaller, dispersed schools are often preferred over large schools in huge camps to minimize children's risk on their route to school. During site selection and delineation, it is essential to consider the possibility that more classrooms will be required in the future. Schools should be positioned in a central area within the camp, with several routes to ensure everyone has safe access (Fig. 5.3) (Tyrer and Fazel 2014).

Early non-formal educational and recreational activities should be developed into a single, unified primary school system within 6 months of the first significant displacement, based preferably on the curriculum of the nation of origin. If the educational system in the nation of asylum is comparable to that in the country or region of origin, and refugee numbers are restricted, resources may be supplied to local schools to allow them to absorb refugee pupils, if this is cost-effective. The refugees and the officials representing the Ministry of Education should be included in the decision-making process (United Nations High Commissioner for Refugees 2015).

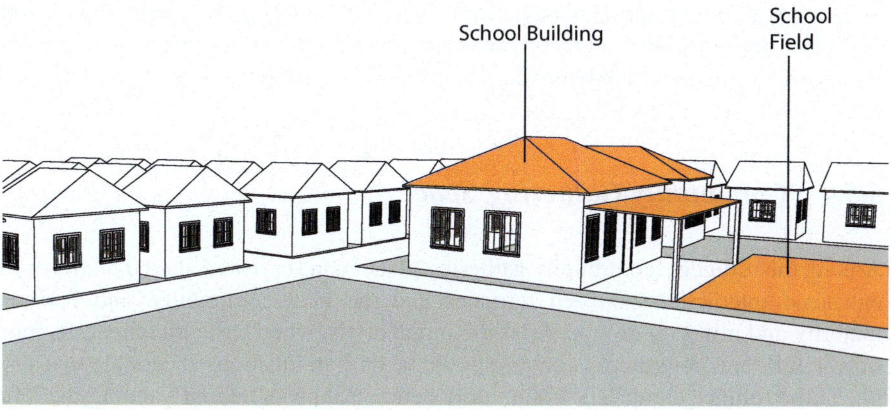

Fig. 5.3 Schools should be positioned in a convenient location inside the camp

5.5.2 Place of Worship

Despite the semi-permanent duration of shelter, rarely is any kind of religious space found in refugee camps. As a long-term need, places for worship can play a vital role in the mental health of the refugees. Places for worship can not only help relieve stress experienced by refugees but also assist in building a strong community. Hence, refugee camps should accommodate religious spaces of worship that will help create a strong relationship to their normal lives (Tyrer and Fazel 2014).

A building structure dedicated to worship can be a semi-permanent type, one that is easy to assemble religious spaces to accommodate the basic requirements of a particular religion. A demographic survey regarding the common religions of the refugees is necessary to determine the type of religious spaces in the refugee camp. While camp planning, it is highly significant to include places of worship for minority groups of refugees as well. Those forced to flee their homes often find consolation and cultural support in places of worship. (Global Shelter Clusters 2018).

5.5.3 Commodity Distribution Centre

To ensure safety and equitable resource sharing, each distribution centre storing and providing goods must serve a reasonable number of refugees. The safe storage of commodities and proper control over their distribution means that distribution centres must be adequate, well-maintained, and secure, must use standardized measures and weigh scales that are properly used by distribution staff and must be trained in the principles and practice of distribution. A safe, covered, sheltered storage room with adequate capacity to retain goods arriving in the camp for many days before distribution should be included in distribution centres (UNHCR 1997).

Standard warehouse norms must be followed, particularly those addressing cleanliness, security, stacking, and record keeping. It is also important that there is a distribution area through which beneficiaries must travel promptly (UNHCR 1989).

5.5.4 Community Gathering Space

Among the refugees, community gathering places can serve as a shared platform for enhanced interaction between refugees and the host community, and between majority and minority groups. After the initial crisis, when basic human needs must first be fulfilled, community centres should be built in locations accessible to everyone in the refugee camp. The community centre can promote unity among the refugees, helping showcase local artists' culture and craftsmanship as a community of creative individuals. Refugee community members can make their items and sell them in the community centre. Children may also utilize this space to study, play, and engage in various activities all in one place (Global Shelter Clusters 2018).

5.5.5 Administrative Buildings

Traditional structures for administrative and communal functions should be used, and if possible, multi-purpose designs should be used to allow for various uses. For example, buildings first utilized for emergency services could eventually be converted into schools or other community facilities. Below is a list of the most often requested administrative and municipal services. The division of categories is primarily for demonstration purposes as the importance of maximal decentralization has previously been emphasized. Administrative and other facilities, whether centralized or decentralized, should be located and structured so they are accessible to everybody (United Nations High Commissioner for Refugees 2015).

5.5.6 Marketplace

Marketplaces should be included in the refugee camp as they can create a long-term livelihood for the refugees living there. Therefore, it is necessary to identify people's informal livelihood skills and capacities, and to support those that the community considers viable, as well as to identify local markets. It is important to ensure that women are not overloaded with additional chores related to re-establishing livelihoods, as women frequently perform numerous roles due to expectations around gendered labour. However, if necessary, allow them to express themselves by holding group sessions (Dalal et al. 2018).

During the day, central marketplaces in camps can be very bustling and crowded. Although there is a lower probability of serious physical attacks, there remains a risk of strangers using the anonymity of vast groups. Central markets may be a long way from many of the shelter clusters. As a result, they may be abandoned at night, potentially becoming either gloomy no-go zones or more likely to be frequented by groups of males, raising the danger of attack or harassment for women. One keyway to reduce the overall risk associated with central markets is to allow the establishment of local stalls, shops, and markets within the shelter sections, as those who own the local stalls are likely to come from nearby shelters and may also be present in those stalls for longer hours in the evenings. While shaded areas or stall-building materials may be required in some camp conditions, it does not have to be complicated. Often, all that is required is to provide appropriate space, in tiny open areas or squares at critical crossing places among the shelter sections (Global Shelter Clusters 2018) (Fig. 5.4).

5.5.7 *Healthcare Centre*

Each refugee community (about 10,000 people) should have its own health centre. Larger settlements may require many clinics. The health centre should be a simple structure with consultation rooms, basic curative care (drugs from the New Emergency Health Kit), oral rehydration therapy, clinical procedures such as dressings (but not injections due to the risk of HIV transmission), a small lock-up pharmacy, simple equipment and sanitization facilities (electricity may not be available), and data collection (logbooks to record patients and activities). In all health facilities, water, sanitation, and equipment for universal precautions are required. Each population of about 50,000 people should have a central health centre with 24-hour

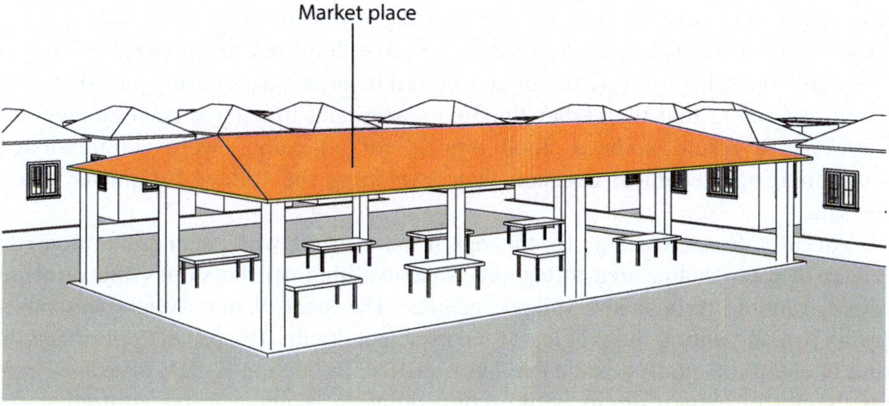

Fig. 5.4 Marketplaces should be located in a secure location accessible to everyone living in the community

inpatient and outpatient services. A proper infection control system must be in place. Basic laboratory facilities may be available. The central health facility should also oversee and train employees (at both the first and second levels) as well as coordinate the primary health programs (reproductive health) (UNHCR 2007).

5.6 Case Study: Rohingya Refugee Camp, Bangladesh

In 2018, two planned camps, accommodating 1340 and 995 homes, were established to minimize congestion in Kutupalong-Balukhali refugee camp in Bangladesh. Beginning in 2019, the project team continued to expand the second camp. The project team demonstrated the low flood risk in the valley sections. With sustainable site development works, the camp's capacity could be increased by more than 40% with minimal environmental impact.

The Bangladeshi government enlarged the perimeter of the largest Rohingya refugee camp in Cox Bazar district in early 2018 to make room for new arrivals and allow families to transfer from the camp's most congested and high-risk regions. Due to the difficult topography of this area, which consists of steep, tightly interlaced hills with almost no flat places, considerable earthworks were required to provide safe areas for shelter construction. However, by 2019, the government had forbidden cut-and-fill interventions, which meant new shelter complexes had to be built in the camp's abandoned areas.

New shelters and site plan layout were developed utilizing a settlement approach, according to the mid-term shelter strategy, to guarantee that the community's overall requirements are satisfied. Access and drainage networks, shelter and WASH layouts, and space for community amenities and open areas for recreation and community gardens were all laid out on site designs.

The project was designed to embrace a "whole settlement approach," incorporating site planning, site development, shelter, WASH, and site management. By 2019, the valley floor was the sole site for construction in this part of the camp. These places were flood-prone, according to the 2018 flood risk map. Despite times of very high rainfall, no severe flooding occurred in these places during the 2018 monsoon, indicating that the earlier flooding assessment may have been unduly cautious. As a result, updated flood models for all camps were commissioned, confirming the technical decision that developing the valleys for shelter would be safe.

An early drainage design for the region was created, with catch drains along the edges of each shelter area to capture water washing off the slopes and principal drains running through the valleys' centres. The level of the shelters was raised using soil dug during the digging of the principal drains. The project prioritized the use of environmentally sustainable Disaster Risk Reduction (DRR) measures, such as the use of natural drains with earth bedding to promote water infiltration and reduce flood risks for downstream communities, and the planting of fast-growing, deep-rooted grasses along embankments and slopes to prevent erosion. In addition, countless people carried out massive tree planting and reforestation initiatives around the camp to restore the ecosystem, protect the slopes from erosion, and decrease floods. This demonstrates the heavy involvement of the refugee community in the planning and implementation to mitigate flooding and environmental disasters (Global Shelter Cluster 2021).

The location and width of pathways, the segregation, and placement of latrines, bathing places, and water stations, as well as street lighting and consideration of generally male-dominated spaces, were all factors considered during site development. The project created approximately 500 mid-term shelter plots in these valleys, boosting the camp's capacity by 40% without needing extensive earthworks and reducing environmental effects (Figs. 5.5 and 5.6).

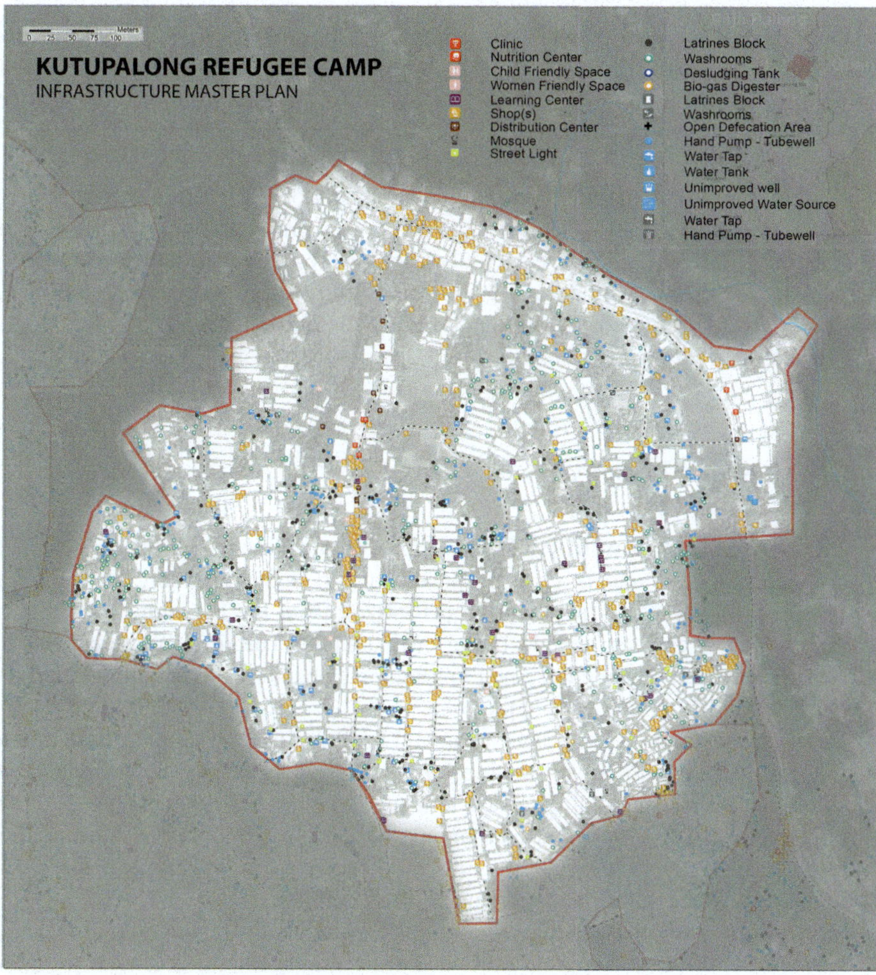

Fig. 5.5 Infrastructure master plan for Rohingya refugee camp in Bangladesh

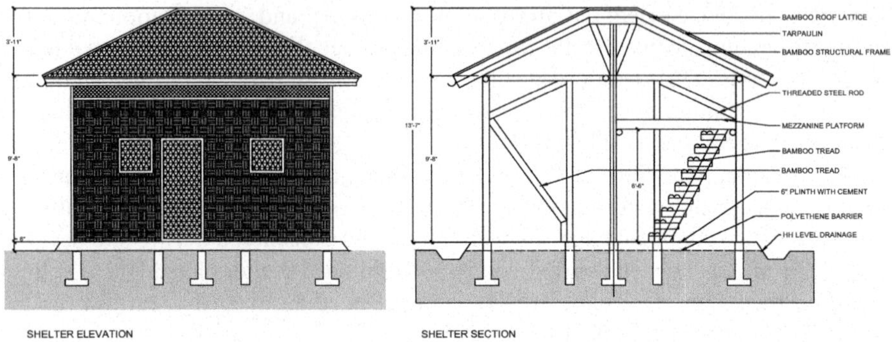

Fig. 5.6 Shelter design (elevation and section)

Once the site preparation work was completed, the site planning team designated the assigned locations for WASH and shelter blocks, which were subsequently built in parallel. Local contractors built the WASH facilities (tube wells, gender-segregated latrines, and bathing places). The plinths of the shelters were raised using soil dug from the soak pits and toilet pits.

Minimizing any negative environmental impacts of the project was a priority consideration at all stages. Therefore, priority was given to protecting what vegetation remained after previous years' deforestation, respecting existing community gardens in site plans, planting alongside drains and on exposed slopes, and using natural drains to promote infiltration and reduce discharge rates and possible flooding downstream (Global Shelter Cluster 2021).

5.6.1 Lessons Learned

- **Infrastructure and utilities**: Proper infrastructure, including drainage, water, and sanitary systems, is critical for the health and well-being of refugees. Planning for these utilities must be prioritized from the beginning to ensure a safe, sanitary, and sustainable living environment. The infrastructure must be semi-permanent, designed to adapt to long-term needs while accommodating the unpredictability of refugee populations.
- **Drainage systems**: An efficient storm and local drainage system are essential to prevent flooding and erosion within refugee camps. Natural water management solutions, such as retaining ponds and properly graded ditches, can be cost-effective alternatives to traditional piped drainage systems. Proper placement of catch basins, manholes, and curving roads reduces maintenance costs while improving water runoff management.
- **Sanitation and waste management**: Sanitation infrastructure, including latrines, solid waste disposal, and wastewater management, is crucial to maintaining hygiene and preventing the spread of diseases. Waste should be managed

to minimize environmental impact, and facilities such as communal latrines and garbage bins should be strategically placed to ensure accessibility and proper waste disposal.

- **Water supply**: Ensuring access to clean water is a fundamental right and must be carefully planned. A reliable water supply system must provide at least 20 L (5.3 gallons) of water per person daily, with sufficient taps, wells, or pumps for the population. Equitable distribution is vital to prevent shortages and ensure all refugees receive adequate water for personal and communal use.
- **Fire breaks and safety**: Fire breaks should be integrated into the camp's layout to prevent the spread of fires. These breaks can double as open spaces for gardening or recreation, enhancing the camp's livability. The placement of shelters should consider the distance required to prevent fire hazards.
- **Community facilities**: Long-term refugee camps must include essential communal infrastructure such as schools, healthcare centres, markets, places of worship, and administrative buildings. These facilities promote social cohesion, community well-being, and economic opportunities for refugees. They also provide access to education, mental health support, and religious practices, which are important for social integration and psychological resilience.
- **Healthcare facilities**: Access to healthcare is crucial in refugee camps. Health centres should be established to serve around 10,000 people each and offer basic curative care, sanitation, and preventive services. Larger camps should have central health centres with 24-hour services and basic laboratory facilities for more extensive care.
- **Social and cultural sensitivity**: Refugee camps must respect cultural, social, and religious practices by providing places of worship, community gathering spaces, and educational facilities. These aspects promote a sense of normalcy, safety, and community for displaced populations, facilitating better mental health and social cohesion.

Weblink of Case Study Project: https://shelterprojects.org/shelterprojects8/ref/A12-bangladesh180821.pdf

5.7 Conclusion

Refugees throughout the world who have been living in host nations for years deserve a better place to live, not simply more functional places. The case study above demonstrates that the refugee community puts their best effort into customizing their camp settings depending on their wants, desires, and dreams. This reaffirms our original stance: a camp should be viewed as an informal city rather than a temporary solution. Refugees significantly remodel their constructed environments via haphazard and unavoidable activities and organization. Therefore, the camp planning should consider their needs for utilities, infrastructure, and community facilities from the beginning.

Discussion Questions

- What are the key benefits of local and storm drainage system?
- Why are sanitary sewers needed?
- What is the best method for water supply distribution?
- What are fire breaks and where they should be introduced?
- What is included in community facilities and what are the social needs?
- What are some challenges that might arise in self-help housing projects, and how can they be addressed?

References

Aburamadan R, Trillo C, Makore BCN (2020) Designing refugees' camps: temporary emergency solutions, or contemporary paradigms of incomplete urban citizenship? Insights from Al Za'atari. City Territ Archit 7(1):12. https://doi.org/10.1186/s40410-020-00120-z

Ajibade OO, Tota-Maharaj K, Clarke B (2017) A simplified guide to surface water drainage systems for refugee camps and internally displaced persons (IDP) temporary settlements. World Environmental and Water Resources Congress 2017: International Perspectives, history and Heritage, Emerging Technologies, and Student Papers-Selected Papers from the World Environmental and Water Resources Congress 2017: pp 519–534. https://doi.org/10.1061/9780784480595.048

Ajibade OO, Tota-Maharaj K, Hills CD, Macleod C (2019) Modelling of a sustainable refugee camp drainage system for stormwater management. Environ Sci (Camb) 5(12):2150–2161. https://doi.org/10.1039/C9EW00350A

Atlas RM (1984) Sanitation and disease—health aspects of excreta and waste-water management. J Am Water Resour Assoc 20(5):803–803. https://doi.org/10.1111/j.1752-1688.1984.tb04765.x

Balbastre-Soldevila R, García-Bartual R, Andrés-Doménech I (2019) A comparison of design storms for urban drainage system applications. Water 11(4):757. https://doi.org/10.3390/W11040757

Cairncross S, Feachem RG (1983) Environmental health engineering in the tropics: an introductory text. No title. Wiley, Chichester

Dalal A, Darweesh A, Misselwitz P, Steigemann A (2018) Planning the ideal refugee camp? A critical interrogation of recent planning innovations in Jordan and Germany. Urban Plan 3(4):64–78. https://doi.org/10.17645/up.v3i4.1726

Ghisleni C (2021) Shelter architecture: the subjective aspects of migrant and refugee settlement projects. In: ArchDaily. https://www.archdaily.com/966887/shelter-architecture-the-subjective-aspects-of-migrant-and-refugee-settlement-projects. Accessed 18 Sep 2024

Global Shelter Cluster (2021) Shelter projects, 8th edn

Global Shelter Clusters (2018) Site planning. Shelter projects

Harvey P (2004) Excreta disposal in emergencies. (October) p 88

Kochary S (2020) Fecal sludge management of duhok city and 4 Idps and refugees camps. J Univ Duhok 23(2):309–320. https://doi.org/10.26682/csjuod.2020.23.2.25

Langar SZ (2020) Beyond refugee housing: 5 examples of social infrastructure for displaced people. In: ArchDaily. https://www.archdaily.com/919496/beyond-refugee-housing-5-examples-of-social-infrastructure-for-displaced-people. Accessed 18 Sep 2024

Sayato Y (1989) WHO guidelines for drinking-water quality. Eisei Kagaku 35(5):307–312. https://doi.org/10.1248/jhs1956.35.307

The UN Refugee Agency (2019) Camp planning standards (planned settlements). UNHCR Emerg Handbook 183(183):1–8

Tyrer RA, Fazel M (2014) School and community-based interventions for refugee and asylum seeking children: a systematic review. PLoS One 9(2):e89359. https://doi.org/10.1371/JOURNAL.PONE.0089359

UNHCR (1989) Supplies and food aid- field handbook. UNHCR

UNHCR (1992) UNHCR manual- water supply in refugee situations. Programme and Technical Support Section, Geneva. https://doi.org/10.4324/9780203850688

UNHCR (2007) Handbook for emergencies united nations high commissioner for refugees, 2nd edn

UNHCR (2021) Education in emergencies—urban. https://emergency.unhcr.org/emergency-assistance/education-and-livelihood/education-emergencies-urban. Accessed 2 Oct 2024

UNHCR (n.d.) UNHCR—The 1951 refugee convention. https://www.unhcr.org/1951-refugee-convention.html. Accessed 17 May 2022

UNHCR TURA (1997) UNHCR commodity distribution: a practical guide for field staff. (June): p 77

United Nations High Commissioner for Refugees (2015) Handbook for emergencies, 4th edn, Geneva

WHO-Geneva, CH, World Health Organization (1988) Urban vector and pest control: eleventh report of the WHO expert committee on vector biology and control, Geneva

Chapter 6
Form and Function of a Shelter

Keywords Refugee shelter design · Shelter typologies · Emergency shelter · Transitional shelter · Durable shelter · Refugee shelter forms · Structural considerations · Refugee camp flood mitigation · Indoor environmental quality for refugee shelter · Safety and security in refugee camp

According to UNHCR (2021), a shelter in the context of a refugee camp is a livable place that provides a safe, healthy, and dignified environment. How a shelter performs will depend on its form and function—the most efficient for its intended use. Apart from life-saving steps, it is a fundamental human right to have a place to live, and this should be maintained by offering shelter and a friendly atmosphere. A shelter's architecture can significantly impact refugee safety and well-being. Moreover, since most refugee crises last much longer than initially expected, this should be considered in the shelter's design from the outset.

This chapter walks the reader into the refugee shelter design criteria. Strategies ranging from conceptual approaches to space design, access, living area, considerations for structure, and human comfort are discussed here. This chapter also delves into the critical relationship between the form and function of shelters, focusing on distinct shelter typologies tailored to the needs of refugees. It classifies refugee shelters into three categories based on their temporary to permanent housing solutions: emergency, transitional, and durable shelters. Emergency shelters address immediate accommodation needs in the wake of conflicts or natural disasters, prioritizing privacy and safety. Transitional shelters bridge the gap between emergencies and more permanent solutions, allowing for gradual adaptation and reuse of components. Durable shelters provide long-term housing, integrating cultural, climatic, and contextual considerations. The chapter emphasizes factors like wind, flood, and seismic resistance alongside indoor environmental quality elements such as natural lighting, ventilation, thermal comfort, personal security, and fire safety. A case study by Balbek Bureau showcases innovative modular town systems designed to provide dignified living spaces for internally displaced Ukrainians, demonstrating the practical fusion of shelter form and function in addressing humanitarian challenges.

A. Friedman, B. Chaki, *Fundamentals of Planning and Designing Sustainable Post-Disaster Shelters*, https://doi.org/10.1007/978-3-031-83317-5_6

6.1 Shelter Typologies

Refugee shelters can be divided into three types depending on the temporary to permanent solution of housing needs for their occupants: (1) emergency shelter, (2) transitional shelter, and (3) durable shelter. First, refugee shelters are needed for immediate accommodation in the advent of conflict or natural calamities. Emergency shelter settlements are usually implemented to respond to the refugees' instant need for living space. Secondly, refugees are transferred to transitional shelters when a more appropriate one is required as a semi-permanent living solution. Consequently, for long-term shelter needs, refugees are provided with durable shelters for permanent living solutions in the refugee camp or housing.

6.1.1 Emergency Shelter

The United Nations High Commissioner for Refugees (UNHCR) defines an emergency shelter as a livable area that provides a safe and healthy environment with privacy and respect.

Emergency shelter plays a crucial role in shaping the social effects of displacement, directly influencing community cohesion, protection, and well-being. Well-designed shelters that accommodate diverse needs—such as family size, cultural practices, and privacy—can promote social stability and reduce tensions, while inadequate or overcrowded shelters may lead to conflicts and competition for resources. Proper shelter also protects vulnerable groups, such as women and children, from gender-based violence, fostering a safer environment. They also offer protection and security for property and assist in maintaining or re-establishing family togetherness. Individual family shelters are generally favoured over communal housing because they provide the required privacy and psychological comfort, which contribute to better mental health and emotional resilience, whereas poor conditions can exacerbate trauma and anxiety. Ultimately, adaptable and thoughtfully planned shelters can facilitate social integration and community participation, easing the transition from emergency response to long-term solutions.

Utilizing materials and designs that people in need or the local population would ordinarily use is the most effective way to meet emergency shelter requirements. When feasible, displaced communities should be able to construct their shelters with the organizational, material, and monetary support required. This will guarantee that the shelter matches their specific needs, foster a feeling of ownership and independence, and significantly cut costs and construction time. It is essential to remember that the initial shelter design, as well as the site selection and planning, will determine the site's long-term sustainability. Consequently, it is crucial to develop a shelter response that extends beyond the emergency phase and is linked to longer-lasting solutions (UNHCR 2021).

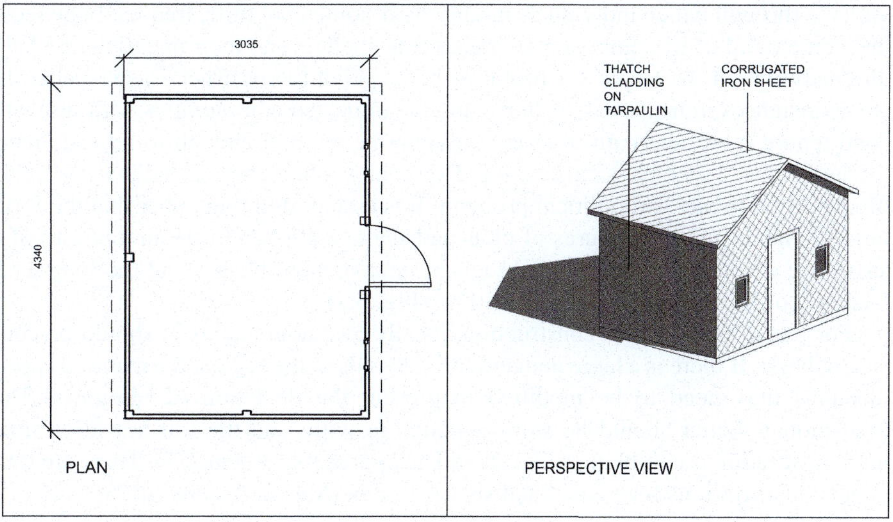

Fig. 6.1 Example of emergency shelter

Emergency shelters are usually built with temporary light building materials that are easy to handle. In many relief initiatives, plastic sheeting (tarpaulin) and thatch have become the most critical component of shelters. Figure 6.1 illustrates an emergency shelter built with tarpaulin or thatch cladding. UV-resistant, heavy-duty plastic sheeting can be used to repair roofs in metropolitan areas (United Nations High Commissioner for Refugees 2015). In some regions, collecting wood from adjacent forests for shelter support frames or stick skeletons can cause significant environmental damage. Therefore, it is essential to provide adequate frame material to support the plastic sheeting constantly. The material for the frame should come from sustainable, renewable sources. Families may benefit from tents, such as when local materials are unavailable or only accessible seasonally, or for refugees with a nomadic past. The longevity of an erected canvas tent relies on its construction, period of storage prior to deployment, environment, and the care of its inhabitants (United Nations High Commissioner for Refugees 2015).

6.1.2 Transitional Shelter

During the period between a conflict or natural catastrophe and the development of a long-term shelter solution, a transitional shelter offers a livable covered living area and a secure, healthy living environment, along with privacy and dignity, to its inhabitants. The term transitional shelter describes a range of shelter choices that assist populations impacted by a humanitarian crisis in transitioning from an initial emergency arrangement to a more acceptable and habitable shelter solution. Often, transitional housing is misinterpreted as a one-time product. In contrast, it is a process through which individuals modify their home over time, frequently by reusing some or all of its components. Transitional shelters and their materials can help in rehabilitation, be resold for profit, or be recycled to construct a new shelter. In some instances, transitional shelters can give impacted communities with insecure land tenure more alternatives since they may be flexible enough to relocate their shelters if faced with physical eviction. However, it cannot be utilized to resolve this frequently intricate and political problem. It is essential to remember that the form of settlements and the features of plots and shelters allotted in the first phases of a crisis are crucial drivers of all subsequent development phases and the long-term viability of the response (Anderson 2019) (Fig. 6.2).

For transitional shelter construction, locally available materials should be considered first. If there is a large demand for materials in the region, the material specifications may need to be modified to prevent the depletion of resources. The transitional shelter should be sturdy enough to withstand the full transition time until restoration is complete, and it should be upgradable and mobile. There are four main considerations for transitional shelter design (Shelter Centre 2012).

- **Upgradable**: While being occupied, a transitional shelter gradually transforms into a permanent housing alternative. This is accomplished by maintenance, expansion, or substituting original materials with more resilient replacements.
- **Reusable**: Transitional dwelling is occupied concurrently with parallel reconstruction efforts. Once the restoration is complete, the temporary shelter can be used for different purposes, such as an outdoor kitchen, barn, or store.
- **Re-sellable**: Transitional housing is occupied concurrently with parallel reconstruction efforts. The transitional shelter is disassembled as soon as rehabilitation is complete, and its components are sold as a resource. Therefore, during the design phase of a transitional shelter, materials appropriate for resale once the shelter has been disassembled must be used.
- **Recyclable**: Recyclable construction materials should be given priority during the shelter material selection. It is necessary to consider that when the transi-

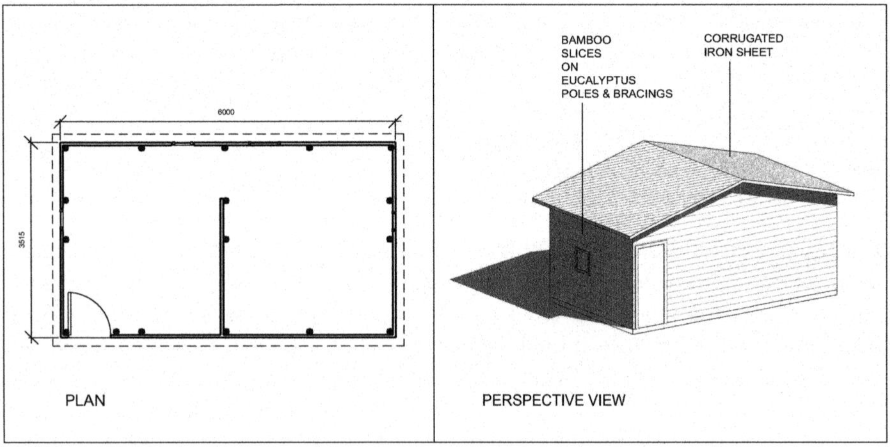

Fig. 6.2 Example of transitional shelter

tional shelter is demolished gradually during the rehabilitation phase, its elements can be used to create a permanent solution for the refugees.

Transitional shelter construction employs basic approaches and speedy building processes suitable for the location and community in which it is implemented. Consideration must be made to how recipients will attain a long-term resolution. This may be accomplished through simultaneous reconstruction efforts or by transforming the temporary shelter into a long-term housing option (Shelter Centre 2009).

6.1.3 Durable Shelter

Although it may be difficult to see beyond the provision of life-saving shelter support during an emergency, it is essential to remember that affected individuals must be assisted in achieving long-term shelter solutions. Within but significantly beyond the emergency and transitional phase, shelters should be altered and contextualized according to climate, cultural practices and habits, local availability of skills, access to suitable building materials, and geographical setting. In addition, requirements for long-lasting shelter solutions should be established following humanitarian standards and in consultation with future shelter users, shelter partners, government agencies, and development partners (Child 2013).

Durable shelters are built with permanent materials (i.e. brick, concrete, etc.). The construction materials should be environmentally friendly, provide thermal comfort and protection from the environment for the long term, and can be obtained sustainably. Using the same materials or kind of shelter as the refugees or local people is the most effective way to meet their lasting shelter needs (UNHCR 2021) (Fig. 6.3).

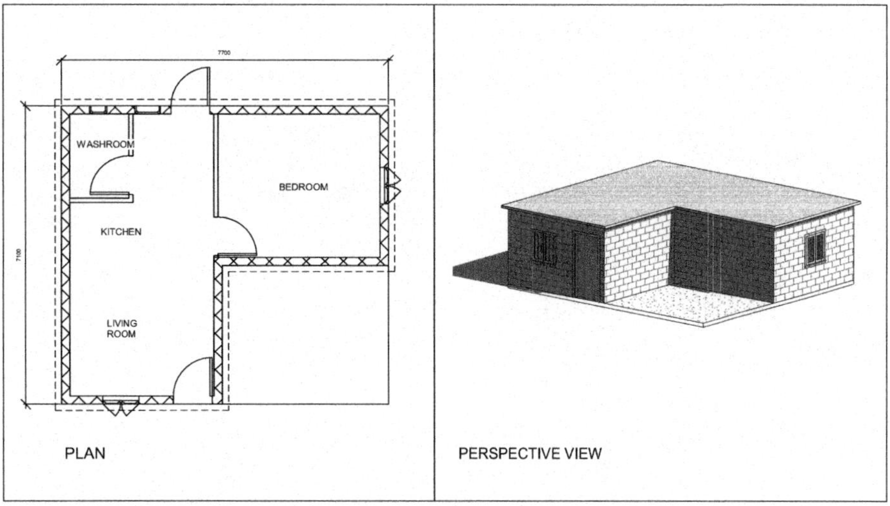

Fig. 6.3 Example of durable shelter

6.2 Types of Dwelling Forms

The shelters can be divided into three types depending on their wall attachment: semi-attached, attached, and row. These dwelling forms have both pros and cons. The designers need to select the dwelling form based on its appropriate use.

6.2.1 *Semi-Detached*

When two structures share a common wall, they are known as semi-detached. The advantage of such an arrangement is in the savings of site area in the host region, providing infrastructure, and avoiding the construction of one wall. In addition, each dwelling will enjoy a sense of privacy and independence, which is apparent in a detached shelter. Greater privacy can be achieved by arranging the units' entrances in different locations. Side entrances will free up more living space and walls for longer windows on the unit's front facade (Fig. 6.4).

6.2.2 *Detached*

Detached homes are the most common type of shelter for refugees. A detached home generally provides privacy, ample outdoor space, and self-contained living. The refugees can use the yard to grow food for themselves. However, many attributes that make a valuable shelter for refugees also contribute to their cost increase. Mounting costs of land and services forced designers to consider narrower lots and homes, which were still detached. The shelters, built on narrow lots, have, as a result, small side setbacks. The home can be placed on the lot line to increase the distance between dwellings (Fig. 6.5).

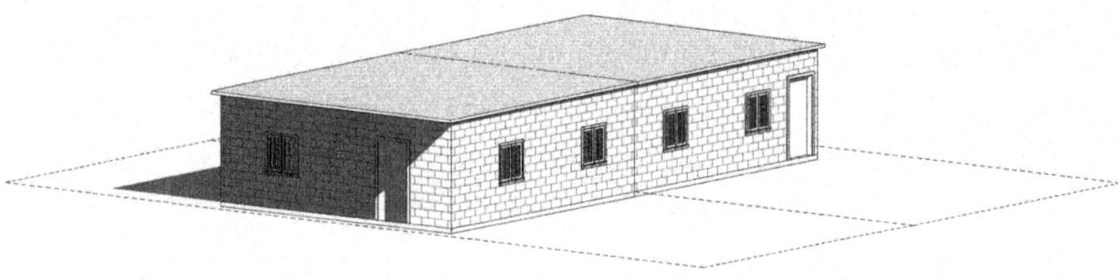

Fig. 6.4 Example of semi-detached shelter

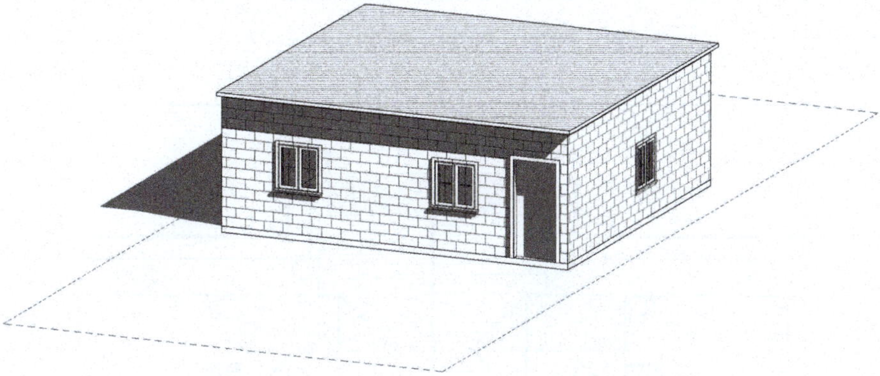

Fig. 6.5 Example of detached shelter

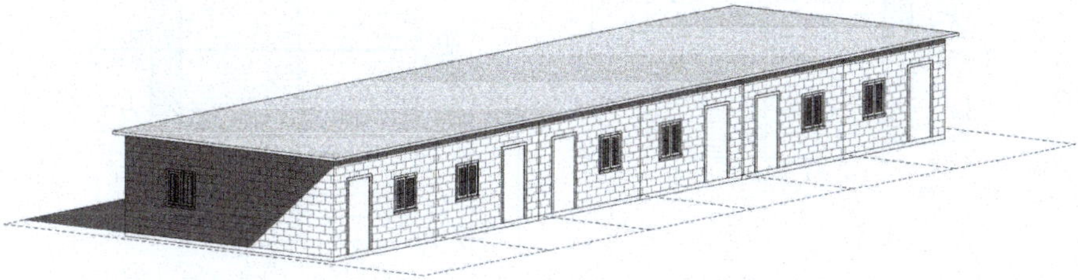

Fig. 6.6 Example of row shelter

6.2.3 *Row*

The building is known as row housing when more than two dwelling units are attached. Building residences in rows can be traced back to ancient settlements. The main advantage of row housing is savings on land and infrastructure costs. The more units are joined together, the greater the savings will be. When designing row housing, a question arises about the preferred number of attached units. Experience shows that joining between 4 and 8 units no wider than 6 m (20 ft.) each will be a recommended number in a row. The decision on how many units to join will also depend on the number of stories, the type of roof, and the street width (Fig. 6.6).

Placing units in a row can be done in a variety of compositions. The objective of the arrangements is to use costly land and infrastructure efficiently, increase livability, and provide opportunities for personalization. One such arrangement is to create a row of interlocked homes. In considering the row's length and composition, a possibility exists to mix housing types. For example, the end units in a row of 6 or 8 shelters can be a multi-family structure, whereas the middle units can be single-family. It will offer a range of units and, as a result, accommodate households with various affordability levels. Another type of row housing is the cluster (Fig. 6.7). Cluster row housing is placed around the open yard, providing each shelter unit with

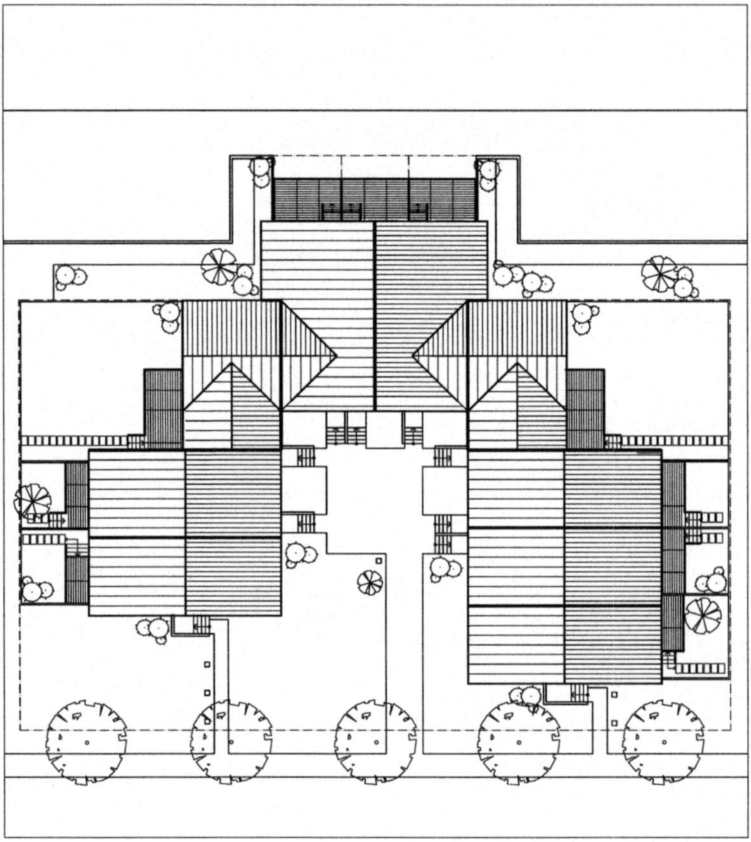

Fig. 6.7 Row shelter cluster

an opportunity for social interaction and enough daylight inside the building. The case study of RE-Ukraine (Sect. 6.9) is an example of a cluster-type shelter. The design utilized cluster-type refugee housing to promote social interaction among displaced individuals. Open spaces within clusters helped to serve as gathering spaces.

6.3 Footprint

The living area for shelters should be sufficient for everyday tasks, including sleeping, cooking and eating, and storing food and valuables. Any shelter should provide a minimum covered living area of 3.5 sq. m. (37.6 sq. ft.) per person in hot climate circumstances and 4.5 sq. m. (48.4 sq. ft.) per person in cold climate conditions (when individuals are compelled to remain within the shelter for longer durations)

(United Nations High Commissioner for Refugees 2015). This criterion can be modified in various circumstances to suit cultural and societal norms. Land shortages, host community customs, and local housing typologies may make it challenging to enforce humanitarian minimum requirements and exacerbate conflicts between the host community and displaced persons. The minimum living space should reflect cultural and social norms, site context, the phase of the shelter response (emergency, recovery, protracted), local codes, and humanitarian standards. Consequently, before adopting the minimum calculated space, these factors must be carefully considered (European Civil Protection and Humanitarian Aid Operations (ECHO) 2017).

6.4 Functional Requirements

A suitable living environment does not just depend on the suitability of the shelter unit. Access to auxiliary spaces and facilities is also considered, as they promote and preserve health, dignity, and safety and enable everyday activities to be carried out. It is essential to offer dedicated cooking areas and provide ample and suitable restrooms and showers to guarantee speedy, safe, and secure access at all times for all users, including youngsters, the elderly, and pregnant women. Communal toilets should be evaluated as an immediate option and replaced as quickly as feasible with individual toilets. Priority should be given to safe and sustainable excreta and solid waste management, which should be adequately controlled to prevent contamination and health problems. The shelter should also incorporate enough storage space for the dwellers. Cooking space can be placed inside the shelter structure or as a separate structure for a specific number of families (UNHCR 2021).

6.5 Structural Considerations

The shelter structure should be designed to withstand any natural calamities. The structural resistance should, therefore, be evaluated to analyze the response of a shelter type to natural hazards (e.g. winds, floods, and earthquakes) (UNHCR 2021).

6.5.1 Wind Resistance

Wind exerts stresses (inward or outward) on all outside surfaces of shelters. If the exterior cladding of a shelter cannot withstand these forces, the structure should withstand them appropriately. Shelters and structures must anchor structural components and cladding to withstand wind-induced overturning, lifting, and sliding and provide continuous load channels for these forces to the base.

6.5.2 Flood Mitigation

There are several types of floods, and each has a unique impact on how it happens, how it may be predicted, and the damage it does. Shelters need to be designed to consider flooding, and mitigation or adaptation measures must be implemented to minimize catastrophic damage, such as temporary evacuation or relocation. The structure of the building should withstand flooding, and the ground floor height must be higher than the flood level.

6.5.3 Seismic Resistance

Earthquake-resistant constructions are designed to withstand the greatest earthquake with a specified chance of occurring in a particular area. Therefore, it is highly significant to determine if the shelter design is suitable for the implementation area's seismic risk. In general, single-story, lightweight structures, such as emergency shelters, are less susceptible to seismic threats as there is no significant risk of collapse.

6.6 Indoor Environmental Quality

Indoor environmental quality (IEQ) is the relationship between a building's environment and its tenants' health. Several elements, such as illumination, air quality, and humidity, determine IEQ. Refugees frequently worry that their symptoms or diseases are the result of exposure to their living environment. Hence, the following criteria should be considered in the shelter design.

6.6.1 Natural Lighting

Appropriate illumination, including access to natural light, is necessary to support the health and well-being of shelter residents. Daylighting is inextricably tied to a shelter's energy consumption and internal environment. Size and location of window openings should be selected depending on a building's adequacy and security requirements. Indoor lighting conditions that are suitable contribute to the health and well-being of the shelter's occupants, provide a living environment that can boost the learning capacity of school-aged children, and also provide adequate space for carrying out certain livelihood activities. Hence, a rule-of-thumb formula must determine the natural lighting requirements for a given shelter type. Natural lighting is sufficient if the total surface area of all windows and ventilation openings exceeds 10% of the floor area.

6.6.2 *Artificial Lighting*

As required, artificial illumination should be supplied to enhance personal safety within and around the shelter. Individual shelters may access artificial lighting on a site level, where only strategic spaces and communal areas are artificially lighted. Different types of artificial illumination sources exist. In addition to candles, which pose significant fire dangers, and connection to an electricity grid or generator, it is also conceivable and recommended to employ alternative renewable energy lighting sources (solar panels).

6.6.3 *Natural Ventilation*

Natural ventilation aids in maintaining a healthy interior atmosphere, minimizes condensation, and limits the spread of infectious illnesses. It also decreases the impact of smoke and other pollutants that may concentrate inside the shelter, such as from cooking or storing harmful items or materials, such as fuel for cooking or heating.

Using a rule-of-thumb calculation formula, the ventilation requirements of a given shelter type are determined and evaluated. Ventilation is sufficient if the combined surface area of all window apertures exceeds 5% of the total floor area. The rule of thumb was derived from the ASHRAE standard, which mandates a minimum air exchange rate of 35 m^3/h (20.6 cfm) per person to ensure internal comfort.

The shelter's floor-to-ceiling height also plays a significant effect in defining its air quality. The sphere regulations specify a minimum floor-to-ceiling height of 2 m (6 ft.-6 in.) at the shelter's tallest point (2.6 m. or 8 ft.-6 in. in warm climates). Moreover, including upper vent apertures (near the ceiling) will enable the warm air to escape outdoors, which is particularly important in warm and tropical climates. In cold regions, the gases and fumes produced by heating stoves must be effectively expelled from the shelter.

6.6.4 *Thermal Comfort*

If a shelter's thermal comfort is appropriate, its inhabitants are comfortably warm or cold, covered, and dry. The ASHRAE standard for a suitable indoor temperature in a mild climate is around 19.44–27.78 °C (67–82 °F) (ASHRAE 2017). These values can grow in hot climates and decrease in cold climates but should never be below 13 °C (55.4 °F). However, the thermal comfort inside a shelter is dependent not only on the temperature and weather patterns of the surrounding area and the shelter materials (particularly the insulating characteristics of the shelter envelope) but also on the experience of heat and cold by the shelter's inhabitants (Fosas et al. 2018).

In a cold climatic region, it is mandatory to provide adequate thermal comfort for the dwellers in the shelter. Keeping the interior of a shelter at a pleasant temperature of 15–19 °C (59–66.2 °F) is highly dependent on the outside temperature, the style of construction, the quality of the insulation, the building's orientation, and the type and size of the stove. In most cold regions, a stove with a performance of 5–7 kW should be able to heat a place with a floor size of 40–70 sq. m. (430.6–753.5 sq. ft.), depending on the conditions. Climates where cold, wet, and snowy conditions dominate for extended periods (3–5 months) necessitate that people primarily reside inside.

Shelters capable of withstanding frigid weather must be of high quality and are expensive to construct. The following should be considered:

- Structural stability (to bear snow- and wind-loads).
- Wind protection of walls, roofs, doors, and windows.
- Protected and heated kitchens and sanitary facilities.
- Heating and chimney provisions.

It is essential to focus on small living spaces and ensure that cold air may be prevented from entering. This can be accomplished using plastic sheets and sealing tapes to seal the room. Cover the windows and doors with translucent plastic sheeting and staple it to the window and door frames. The living space's walls, ceilings, and flooring should be built to effectively insulate against cold air and retain warm air (United Nations High Commissioner for Refugees 2015).

6.7 Personal Security and Fire Safety

Personal safety is crucial at a shelter, to give protection and help at-risk groups feel secure. This criterion is very context-dependent and may vary from household to household and person to person, as well as by family/household composition differences, such as male or female-headed households, single-parent households, same-sex parental households, or number of children or past domestic conflict. Other factors influencing personal safety include tensions with host communities or with other displaced groups. In this regard, the safety of individuals in a shelter should be assisted by including the following:

- Locks on the inside and outside of all doors and windows.
- Artificial lighting sources.
- Sturdy, non-transparent shelter cladding materials.
- Window openings no larger than 60 by 60 cm (2 ft. by 2 ft.).

Site planning and shelter design should be based on fire risk assessments. Including 30-m fire breaks every 300 m between built-up areas (free of garbage, litter, and other objects at all times) and guaranteeing a minimum gap of 2 m. (6 ft.-6 in.) between structures. Ideally, twice the height of the shelters (to the ridge) will ensure safe evacuation in the event of a fire. The selection of shelter materials, consideration of local cooking customs, and provision of defined cooking areas (on individual plots or communal kitchens) will reduce fire hazards.

On-site and individual fire prevention measures should be taken based on the three mitigation strategies:

- Limiting the spread of fire and the production of smoke.
- Ensuring a safe escape time and safe distances between shelter units and solid evacuation strategies.
- Educating the populace about fire prevention and management.

The difference between fire-resistant and fire-retardant materials is that fire-resistant materials are meant to resist burning and withstand heat. In contrast, fire-retardant materials are designed to burn slowly. Flammable materials, on the other hand, are combustible substances that ignite easily at room temperature and catch fire instantly when exposed to flame (Kazerooni et al. 2016).

6.8 Accessibility and Egress

To accommodate persons with disabilities and those facing mobility or access obstacles (such as the elderly, children, and pregnant women), the shelter must be designed with the following characteristics in mind:

- All shelter entrances must be more comprehensive than 90 cm (3 ft.) (to accommodate wheelchairs).
- If the shelter is high or has steps or other architectural barriers, a ramp with handrails and a parapet should be constructed to aid access.
- The door to the bathroom area should swing outwards.

6.9 Case Study: Ukraine, Balbek Bureau, Kyiv, Ukraine

Since Russia's 2022 invasion of Ukraine commenced, about 7.1 million Ukrainians have been displaced internally. In response to this escalating humanitarian situation, the Kyiv-based firm Balbek Bureau has designed a temporary housing system to provide internally displaced Ukrainians with a dignified domicile. This shelter prototype is designed to adapt to varying terrain types and population densities while being rapidly deployed. While the idea was planned for non-battle zones in Ukraine, the framework can potentially include refugees from abroad.

The project employs a basic 6.6 by 3.3 m (20 ft. by 10 ft.) module that may include private rooms, shared spaces for cooking, sanitary facilities, and common areas. The modules are integrated into bigger, self-sufficient residential structures, which can then be combined with playgrounds and green spaces to form neighbourhoods. The architects say the completed community may house more than 8000 individuals. The modular town system is intended to provide a modicum of dignity for their fellow citizens through the expedited delivery of an adaptable set of structures centred on four fundamental elements: private rooms, public recreation areas, adequate kitchens, and bathrooms that can accommodate the needs of those with infants. The company's primary objective is to preserve a reasonable standard of living for temporarily displaced Ukrainians.

Slava Balbek, architect, founder, and chief executive officer of Balbek Bureau, explains that the shelter type is a modular settlement concept for displaced individuals. The project's objective is to maintain a dignified way of life for displaced individuals. The approach is supposed to be deployed quickly: inhabitants can move into the community 2–3 months after construction begins (Angelopoulou 2022).

As discussed previously, cluster-type refugee housing can promote social interaction, and the open space in the middle can be utilized as a gathering space. The size of a residential block determines all design units' dimensions. This is a universal size, allowing the construction of a town from any available blocks that match particular standards. There is an opportunity to enhance or scale the construction and dwelling volumes at the residential component of the block or master plan without structurally affecting the central block. In other words, engineering blocks and community facilities must be planned and built with the possibility of future residential development in mind. Variability in residential sector typology allows the schematic grid of structures to be adapted to plots of various shapes and topography. The design also enhances the indoor air quality and natural light inside the living area (Figs. 6.8 and 6.9).

The effort has attracted support from government leaders, architects, and businesses. Multiple businesses are prepared to begin production of the construction system, available in three variants: wood frame, SIP panels, and sandwich panels. Simultaneously, the Chernivtsi City Council has announced the construction of a temporary settlement centre under the RE: Ukraine framework and other settlements are currently under consideration in Uzhhorod, Ternopil, Tyachiv, and Kyiv (Angelopoulou 2022).

6.9.1 Lessons Learned

- **Shelter typologies**: Refugee shelters can be categorized into three types—emergency, transitional, and durable shelters. Emergency shelters are temporary, addressing immediate safety and privacy needs, while transitional shelters bridge the gap between emergency and permanent housing. Durable shelters are designed for long-term living, considering cultural, climatic, and contextual factors.
- **Local participation in shelter construction**: Involving displaced communities in building their shelters fosters ownership, independence, and cultural appropriateness, while reducing costs and time. Materials should be sourced locally when possible to ensure sustainability and accessibility.

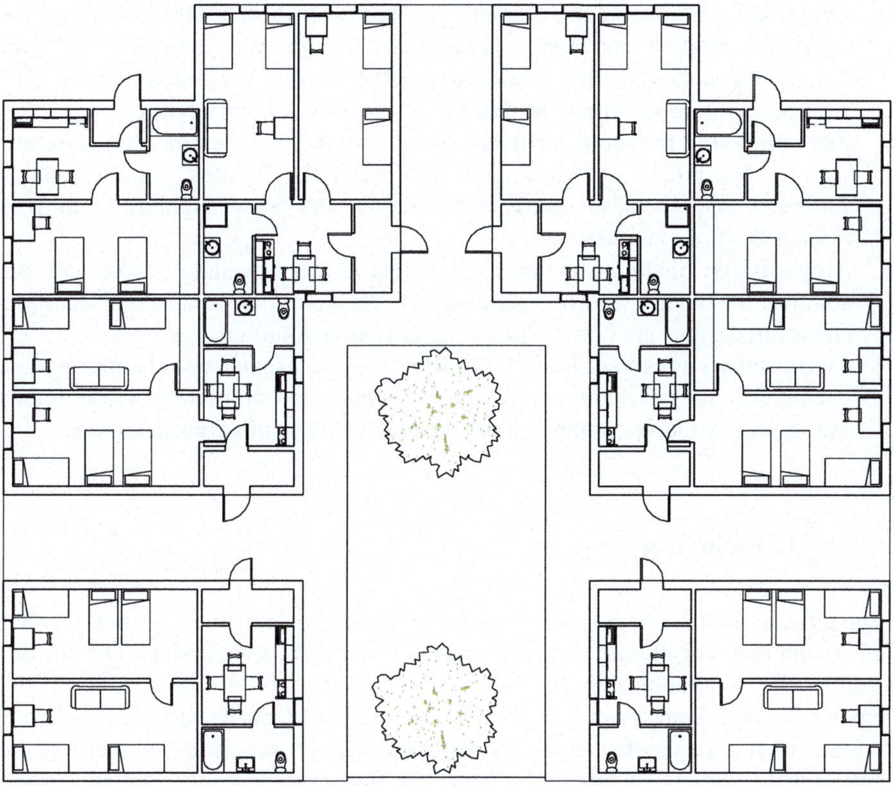

Fig. 6.8 Conceptual plan of refugee shelter prototypes

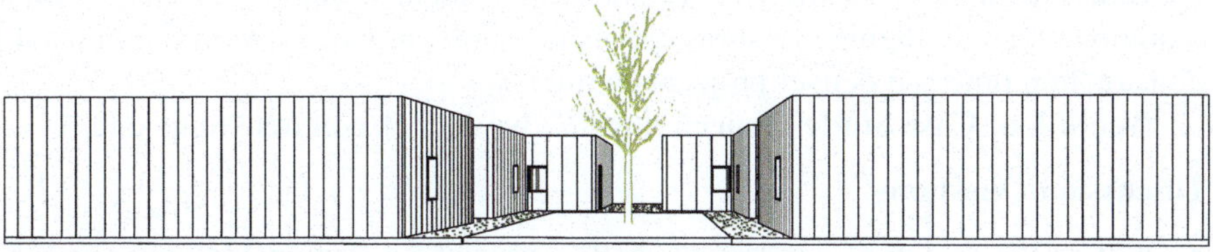

Fig. 6.9 Perspective of refugee shelter prototypes

- **Transitional shelter design**: Transitional shelters should be designed for adaptability, allowing them to be upgraded, reused, resold, or recycled. These shelters are critical in providing stability while refugees transition to more permanent housing solutions.
- **Structural considerations**: Refugee shelters must withstand environmental hazards like wind, floods, and earthquakes. Appropriate materials and design techniques ensure that shelters provide safety and durability in various climates and terrains.

- **Functional requirements**: Beyond the physical shelter, adequate space for cooking, sanitation, and storage is necessary for maintaining hygiene and comfort. Shelters should offer privacy and accommodate the needs of vulnerable groups, including children, the elderly, and persons with disabilities.
- **Fire safety and personal security**: Shelters should include fire safety measures such as fire breaks and non-flammable materials. Personal security can be enhanced through locks, sturdy construction, and proper lighting to mitigate risks of theft or violence.
- **Accessibility**: Shelters should be accessible to all, especially people with disabilities, the elderly, and pregnant women. Door widths, ramps, and accessible bathroom designs are critical for inclusive shelter solutions.
- **Community engagement and livability**: Shelter design should facilitate social interactions through community spaces, while also promoting psychological well-being by incorporating cultural, social, and religious considerations.

6.10 Conclusion

The refugee shelter design is contextual, and there is no universal solution for shelter design that will fit every refugee shelter context. The shelter should be customized to the geographical setting, climate, cultural traditions, and habits. While emergency responses frequently include the supply of tents or emergency shelters, it is essential to ensure that shelter aid programs may grow towards more lasting and sustainable solutions, utilizing local materials, skills, and building techniques whenever feasible. Finally, regarding the shelter design, the designers should consider the permanence of the structure and the fast construction process by the beneficiaries. Considering all the criteria, the prefabricated shelters can be a good option. Additionally, by utilizing prefabricated shelter units, it is possible to minimize the construction time and deliver projects on time.

Weblink of Case Study Project: https://www.balbek.com/reukraine-eng

Discussion Questions

- Why is the design of emergency shelters crucial for sustainable living and the well-being of refugees?
- How do the form and function of a shelter impact its efficiency and suitability for its intended use?
- What are the key considerations for transitioning from lightweight tents to more permanent transitional shelters?
- How should shelters be designed to ensure structural stability and protection from outdoor climates?
- What factors should be considered regarding construction time and cost for emergency shelters?

References

Anderson T (2019) Transitional refugee housing. Unpublished M. Arch Thesis, University of Cincinnati, Cincinnati, OH

Angelopoulou SL (2022) Balbek bureau develops RE: Ukraine, a modular town system for refugees. In: Designboom https://www.designboom.com/architecture/balbek-bureau-re-ukraine-modular-town-system-refugees-03-30-2022/. Accessed 14 May 2022

ASHRAE (2017) ASHRAE technical FAQ. https://www.ashrae.org/File/Library/Technical Resources/Technical FAQs/TC-02.01-FAQ-92.pdf. Accessed 22 May 2022

Child J (2013) Beyond the minimum: a durable emergency shelter. Unpublish M. Arch Thesis, University of Waterloo, Waterloo, Canada

European Civil Protection and Humanitarian Aid Operations (ECHO) (2017) Humanitarian shelter and settlements. https://civil-protection-humanitarian-aid.ec.europa.eu/what/humanitarian-aid/shelter-and-settlements_en. Accessed 30 May 2024

Fosas D, Albadra D, Natarajan S, Coley D (2018) Improving the thermal comfort in new shelters: design recommendations for Azraq refugee camp. https://doi.org/10.6084/M9.FIGSHARE.8977556

Kazerooni Y, Gyedu A, Burnham G, Nwomeh B, Charles A, Mishra B, Kuah SS, Kushner AL, Stewart BT (2016) Fires in refugee and displaced persons settlements: the current situation and opportunities to improve fire prevention and control. Burns 42(5):1036–1046. https://doi.org/10.1016/J.BURNS.2015.11.008

Shelter Centre (2009) Transitional shelter guidelines. Geneva, chrome-extension://efaidnbmnnnibpcajpcglclefindmkaj/https://www.iom.int/sites/g/files/tmzbdl486/files/migrated_files/What-We-Do/docs/Transitional-Shelter-Guidelines.pdf. Accessed 30 May, 2024

Shelter Centre (2012) Transitional Shelter Guidelines. Geneva, chrome-extension://efaidnbmnnnibpcajpcglclefindmkaj/https://www.iom.int/sites/g/files/tmzbdl486/files/our_work/Shelter/documents/Transitional%20Shelter%20Guidelines%20-%202012.pdf. Accessed 30 May 2024

UNHCR (2021) Shelter and Sustainability, chrome-extension://efaidnbmnnnibpcajpcglclefindmkaj/https://emergency.unhcr.org/sites/default/files/2024-01/UNHCR%20shelter%20and%20sustainability%20overview%20April%202021.pdf. Accessed 30 May 2024

United Nations High Commissioner for Refugees (2015) Handbook for emergencies, 4th edn, Geneva

Chapter 7
Prefabricated Shelters

Keywords Prefabricated shelters · Emergency housing · Self-help construction · Prefabricated shelter prototypes · Modular refugee housing · Kit shelter · Low-cost refugee shelter · Refugee shelter design · Refugee communities · Shelter construction methods

Prefabricated dwelling units are commonly used as an effective solution for emergency shelters. Prefabricated prototypes, mostly ready-made structures, are shipped to the disaster area during the emergency, either in pieces or fully assembled, built, or placed on-site. Prefabrication can also ease the self-help construction process. Prefabricated self-help kit homes are a potential solution to reduce housing costs and allow refugees to participate in the decision-making process concerning the design and build of their own homes. This is possible by eliminating paid labour and replacing it with sweat equity, allowing the shelter dwellers to have a high level of involvement during the pre-occupancy and construction stage.

The content of this chapter encompasses various aspects of prefabrication methods, advantages, disadvantages, and specific examples. Prefab shelters are constructed in a factory and then assembled on-site, often featuring roofs, steel or aluminium frames, and walls made of polyvinyl or glass. They offer benefits like faster construction, higher quality workmanship, weather protection during manufacturing, and reduced labour costs. The shelters can be used for emergency or permanent housing and are suitable for various environments. Different materials and methods, such as modular, panelized, and kit-of-parts approaches, are employed in creating these shelters. A case study related to the Tohoku Earthquake in Japan illustrates the use of prefab shelters in disaster-stricken areas, covering various aspects of construction, challenges, and site layouts.

A. Friedman, B. Chaki, *Fundamentals of Planning and Designing Sustainable Post-Disaster Shelters*, https://doi.org/10.1007/978-3-031-83317-5_7

7.1 Prefabrication: The Fundamentals

Prefab shelters are buildings erected on-site after being partially or entirely prefabricated in a factory. Most prefabricated structures consist of a roof and a steel or aluminium frame. Some shelters feature polyvinyl or glass walls as protection from the outdoors. Prefab shelters can be built faster than conventional construction, saving the contractor money on labour and material expenses. They also offer durability and minimal maintenance while withstanding challenging environments like high humidity and temperatures (Fig. 7.1).

7.1.1 Advantages of Prefabrication Methods

Prefabrication methods offer many advantages when compared to traditional, on-site construction. The rising trend in prefabrication can be attributed to its financial, environmental, structural, and time-saving advantages. Building components assembled in a controlled factory setting are protected from harmful weather conditions and tend to have higher quality workmanship than site-assembled ones. Moreover, constructing houses in a factory negates problems such as vandalism, material storage, and weather delays. Factories also require fewer skilled workers, making the end product cheaper. An efficient, high-tech factory can produce a house in a week, compared to on-site construction, which takes over 5 months on average. Product testing also ensures that they meet the manufacturer's quality standards. Furthermore, reducing on-site construction time reduces labour costs and limits damage due to warping, rot, vandalism, and theft. The assembly of prefabricated

Fig. 7.1 Prefabricated refugee accommodation or asylum seekers hostel in Germany

parts usually requires a smaller team of workers, and construction coordination is simplified as a result (Lopez and Froese 2016) (Fig. 7.2).

Prefab shelters may meet any emergency shelter requirements through design and engineering. Prefab shelters, usually movable buildings, can be used as temporary or permanent structures; therefore, they are highly recommended for emergencies. They may be utilized indoors and outside and were created as a component of an existing building system. The prefab shelter typically consists of a frame and panels, although windows or shelves can be added to make it specifically meet the use. Customizing a prefabricated building's design may increase costs, although such modifications in certain circumstances could be beneficial.

Aluminium, precast concrete, wood, and steel are some materials that may be used to build prefabricated shelter structures. The case study of Prefab Housing in Japan (Sect. 7.4) used a mix of prefab light metallic structures and wooden units. The steel structures were more prevalent due to quick assembly. Prefab shelters, both their whole and their parts, are produced in regulated factory settings. The completed prefab structure or sections is transported to sites prepared for construction. All required elements, including walls, floor trusses, windows, heating and cooling, plumbing, electrical wiring, and interior finishes, are included in prefabricated shelter kits. Manufacturers of prefabricated shelters provide flexibility and benefits over the on-site building including, factory-controlled quality, and ease of extension and relocation (Gupta 2012).

As is the case with any manufactured component, waste generated by the prefabrication of panel systems is less than what could be expected from on-site construction. Additionally, since prefab shelters are constructed in factories rather than on-site, there is no material waste. After building one prefabricated shelter, everything left over may be utilized to build another one in the factory (Roberts and Writer 2020).

One of the most significant advantages of prefabricated panel systems is the superior quality that can be achieved through the manufacturing process. This quality can be evaluated based on three interrelated characteristics: craftsmanship, technical performance, and durability. The system's craftsmanship governs its potential to achieve consistent performance levels from one application to another. Construction management and trade coordination can be simplified with these products, and the need for large teams of skilled, on-site labour for multiple-unit construction is substantially lowered (Roberts and Writer 2020).

Unsheathed structural panels appear to perform well in all respects but benefit from a few extraordinary characteristics. The panels' biggest advantage is overcoming the inadequate workmanship found in conventional construction without resorting to very unfamiliar building techniques.

The large-scale repetition of standardized building components significantly reduces manufacturing and installation costs. Furthermore, the small-sized refugee shelter design allows for the easy transportation of building components (Chauhan et al. 2022).

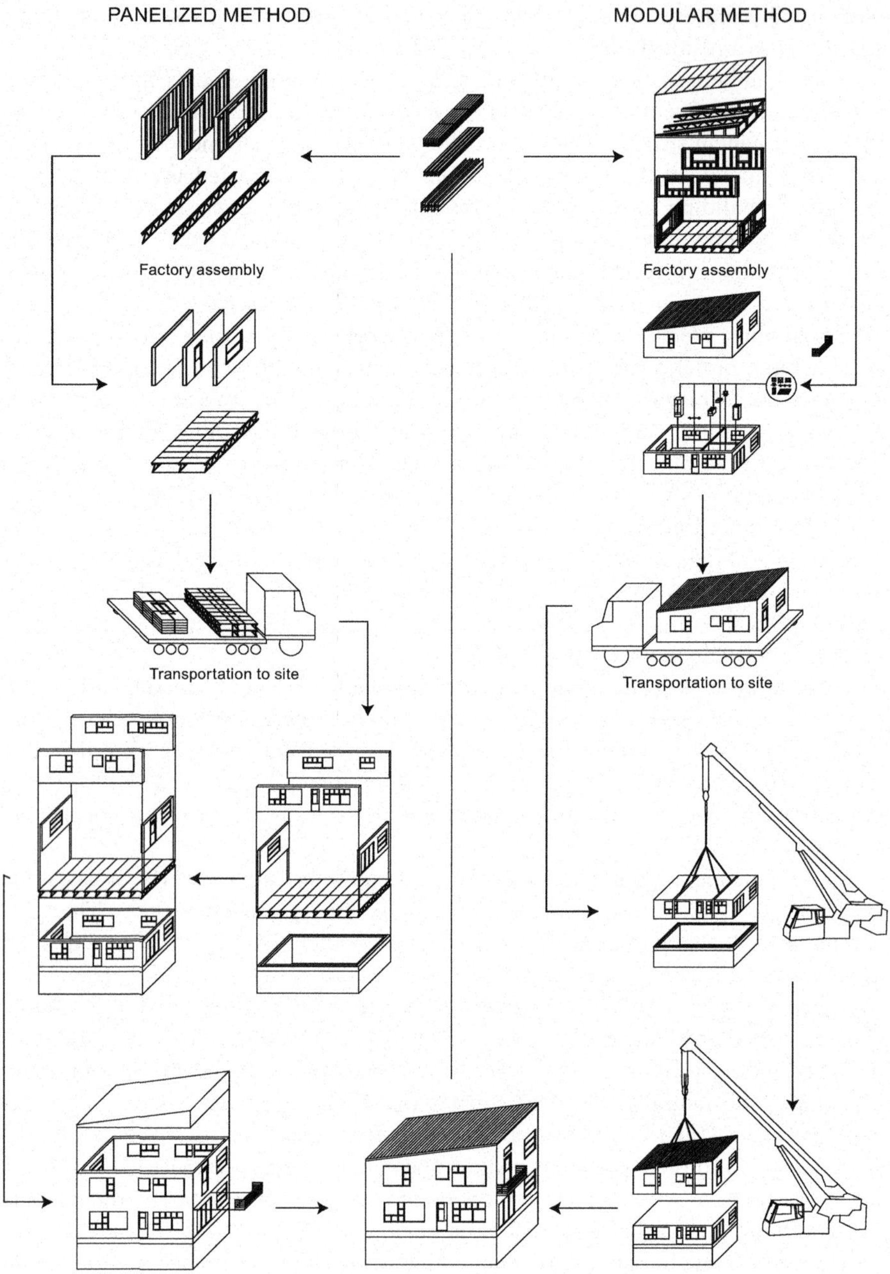

Fig. 7.2 Steps leading to the construction of a townhouse using panelized and modular prefabrication systems

7.1.2 Disadvantages of Prefabricated Shelter

Prefabricated construction methods, specifically designed emergency shelter units, and even winterized shelter units have not successfully housed refugees. This is because prefabricated shelter units are more expensive than regular shelter construction. Moreover, producing prefabricated shelter kits and shipping time to the emergency sites sometimes take a long time. Transportation of shelter units to the construction site can also add to the overall construction cost. Prefabricated shelter design should consider the cultural and social norms of the community in emergency needs. Otherwise, a prefabricated shelter might not be the best fit for the emergency site. Lastly, cooling is a regular problem for conventional prefabricated shelter designs for emergency housing based on hot climatic regions. The common disadvantages of prefab refugee shelters must be considered while designing and choosing shelter prototypes for emergency housing (UNHCR 2024).

7.1.3 Main Types of Prefabrication Methods

There are several types of prefabricated systems, subsystems, and components that can be combined at various levels to provide a complete system package.

Modular prefabrication refers to factory-built, nearly complete dwelling sections. An entire level of a house is manufactured, shipped to the site, and positioned on foundations using a crane. Using this method, a row of shelters can be quickly built up as a series of stacked modular blocks. The main advantage of modular prefabrication is the speed at which dwellings are erected. One needs to be cognizant that the delivery and the installation process undermine savings gained by the construction efficiency.

Panelized prefabrication produces wall panels of varying sizes and stages of completion in a factory either manually or using fully automated methods. Manufactured panels are stacked in trucks, shipped to the site, and assembled to become the exterior walls. The walls are significantly more airtight than on-site framed ones. Exceptionally well-fitted joints create a continuous thermal break and provide excellent insulation values for their thickness. Airtightness will prevent condensation from forming in the walls and increase the product's durability. Panelized methods are the most used prefabricated since they are similar to traditional on-site building practices.

Kit of parts refers to a collection of separate building components that are pre-engineered and intended to be combined in various ways to form a finished building. In the "kit of parts" prefabrication system, well-marked house components such as studs, floor or roof trusses, doors, windows, stairs, and exterior finishes are manufactured and delivered to the site where they are assembled and installed. In kit-of-parts "Catalogue Design," the designers prepare the catalogue for various house designs, and the owners select from the catalogue according to their choice.

A kit-of-parts catalogue design approach utilizing prefabricated building components is especially suited for townhouse construction. For example, using a kit of parts, prefabricated dormers, bay windows, and entrance overhangs can be offered to home buyers who wish to customize their façade (Friedman 2021).

7.1.4 Choice of Method

Panelized systems are also suited for rowhouse-type shelter construction since their narrow footprint requires smaller panels, facilitating transportation, delivery, and installation. A greater number of panels may increase production costs and assembly time, yet they will offer greater exterior variation and avoid monotony. Therefore, prefabricated shelters consisting of a lower number of panels are most suitable for emergency shelters. However, prefabricated shelter kits with more panels can be chosen for durable shelter types (Fig. 7.3).

From uninsulated to nearly complete walls, a wide variety of prefabricated panel options are available to builders. The number of shelters to be constructed and the time frame will determine the selection of an appropriate panel system. Open-sheathed panels are commonly built with 5 cm by 15 cm (2 by 6-in.) studs with either plywood or aspenite sheathing or with 5 cm by 10 cm (2 by 4-in.) studs with

Fig. 7.3 Automated manufacturing process of prefabricated panels

extruded polystyrene sheathing. Panels are left open to the interior to install insulation, electrical wiring, and plumbing on-site. In structural sandwich panels or foam-core panels, a core of rigid foam insulation is laminated between two facing materials. Panels can be sold with additional integrated exterior or interior finishes. Unsheathed structural or composite panels are built using a combination of wood or metal structural elements with rigid foam insulation infill. Four structural combinations are available, each incorporating a larger portion of the building envelope. A nearly complete panel will have, for example, exterior air barriers and interior finishes (Keena and Friedman 2024).

Interior partitions can also be prefabricated. The option to easily add or remove partitions will greatly increase a townhouse's adaptability. Demountable partitions allow dwellers to make post-occupancy modifications to suit their changing space needs (Friedman 2021).

7.1.5 Plug and Play Type Prefabricated Dwelling

"Plug and Play" homes are prefabricated, modular units that can be installed rapidly and made ready for immediate use. In principle, with modern shipping techniques, Plug and Play dwellings can be transported or relocated worldwide. Unlike prefabricated houses, which require some on-site assembly and are typically permanent, Plug and Play houses require no assembly upon delivery. They simply need to be connected to utilities once placed on-site. The low associated costs, simplicity, and ability for transportation make Plug and Play dwellings an exciting new housing type. Plug and Play shelter designers need to pay special attention to two aspects affecting their conception: transportation and interior design features (Friedman 2021).

The design of Plug and Play units is determined by their need to be easily transported without causing external or internal damage. During transit, weight becomes a crucial factor that permeates many other design decisions. The goal is to create heavy units to sustain maximum wind drag when erected but not too heavy to exceed the supporting allowance on rooftops. For such rooftop units, the weight varies greatly according to units and material choices (Fig. 7.4).

Designing to facilitate transportation involves protecting external building components. Windows and doors present a unique challenge since they weaken the structure and are quite breakable. Rather than compromising and limiting the number of windows and doors, designers have developed methods to protect them during travel. One method consists of sliding large protective panels over the windows. This allows the design of Plug and Play units in which the facades are almost entirely glass. Furthermore, some designers have created units where the doors and windows fold into the surrounding structure to protect them during transport. This also provides for more interior space once the unit is installed. It is also possible to use thick shutters, which can be closed to seal off the glass and add structural support, much like hurricane-proof structures. Designers have also suggested units that

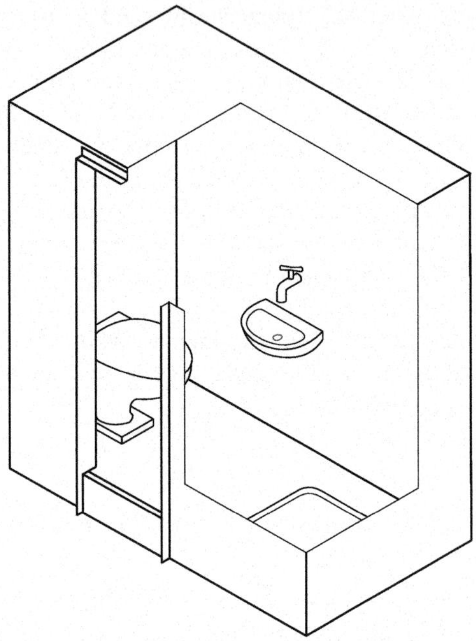

Fig. 7.4 Example of Plug and Play washroom unit

can mechanically retract inward to protect them during transportation and then be extended back out at the push of a button (Shaw 2015).

Since Plug and Play shelters are generally small, designers must consider interior design for a limited space. The goal is to create spaces that feel large while still being efficient and functional. While this includes many concepts covered in the chapter on small-sized shelters for emergency needs, several principles are uniquely relevant to Plug and Play homes. Furthermore, assumption exists that modular, transportable units are cheap and uncomfortable. While this may have been true, recent innovations and technology have created high-tech, aesthetically pleasing structures that are sleek, efficient, and lightweight (Melis 2003). Because Plug and Play shelters are smaller and less costly than traditional shelters, designers can use sophisticated technologies and customizable features to make them easy and fast to construct and well suited to shelter needs.

The idea of built-ins can also be extended to the arrangement of utilities. To save space and minimize clutter, some designers have implemented heating and cooling on the floors and walls. Furthermore, the shelters are typically designed so that the windows, or roof, can open to let outdoor air circulate throughout the unit. This can function as natural air-conditioning and provide significant cooling effects.

7.2 Prefabricated Homes for Refugee Communities

7.2.1 Prefabricated Container Shelters

Innovative flat-pack prefabricated shelters provided with good insulation and designed for quick assembly have been proved as one of the best solutions for refugee camps by responding immediately to the significant demand with huge production capacity. To address the housing shortage, prefab buildings quickly supplied dwellings for refugees in Turkey, particularly those from Lebanon, Germany, and Syria. The refugee container became a crucial accommodation option in natural disasters, including the 2010 Haiti earthquake, the 2011 Turkey Van earthquake, and the 2013 Pakistan flood disaster. The considerable demand for refugee shelters following war, disaster, or persecution can be met by offering prefabricated modular shelters that are mass-produced and brought ready to the refugee camp (Karmod n.d.).

Many manufacturers of prefabricated structures have come forward with the supply of modular emergency refugee shelters with an extensive production capacity reaching up to 1100 containers per month. Shelter organizations in Syria following the war have successfully provided shelter to refugee communities by delivering prefabricated modular container shelters in a very short time to accommodate thousands of refugees in the refugee camps (Fig. 7.5).

These prefabricated container refugee shelters have a bathroom, shower, kitchen cabinet, one or two rooms, lights, and outlets—everything a family needs to live comfortably. Depending on the situation, air conditioners, water tanks, window

Fig. 7.5 Prefabricated container shelter for refugees

coverings, shutters, and window burglar bars can all be added. Prefab shipping containers and polyester cabins are the most practical, affordable, and rapid ways to construct refugee camps that can house thousands of people quickly.

The prefab containers are created with a comfortable and contemporary style and are in accordance with global transportation regulations, making them simple to move. Once they are on-site, they are very simple to build. Prefab containers are available in various sizes and functional layouts, including 2.3 by 6 m. (6.9 by 18 ft.) 3 by 7 m. (10 by 21 ft.),3 by 9 m. (10 by 27 ft.), and 3 by 12 m (9 by 36 ft.). The inside layout may be modified to accommodate a single, large, or even two families. The demountable modular container structure enables numerous units to be connected to create a variety of facilities, including schools, healthcare centres, clinics, places of worship, laundry rooms, play areas, restrooms and showers, dining halls, and other social facilities needed inside any refugee camp (Karmod n.d.) (Fig. 7.6).

The containers are made entirely of materials selected for their extended lifespan. After the emergency that forced refugees to flee their homes has ended, the containers may be conveniently stacked for future use in emergencies anywhere in the world (Karmod n.d.).

7.2.2 *Ikea's Prefabricated Relief Shelter Unit*

Ikea, the Scandinavian maker of assemble-it-yourself furniture, introduced the Better Shelter project to aid the world's rising refugee crisis. The Better Shelter initiative started as a design research project in the small town of Hällefors in Sweden in 2009, which is formally a part of the business' charity division, aimed to create readily erected shelters that were not just more robust or heavy-duty but also less expensive. The final product, of which 30,000 have already been constructed and are in use, offers a workable answer for the more than 65 million refugees who are now homeless globally (The Guardian 2017).

The prefabricated shelter unit has a lockable door, 4 windows, and 4 ventilations. The shelter is made of a semi-rigid opaque roof and wall panels. The structure is made of a steel frame that can be mounted with ground anchors. It can be outfitted with solar power-generating photovoltaic systems and portable lamps. The flat-packed shelter has 5 years of shelf life, a minimum of 3 years lifespan for panels, and a minimum of 10 years for the frame. Constructing the shelter can take as little as 4–6 h of assembly by a team of 4 persons. It is possible to assemble 140 shelter units daily (BetterShelter n.d.).

The size of the shelter is 5.68 m (18.6 ft.) × 3.32 m (10.9 ft.), having an area of 17.5 sq. m (188.4 sq. ft.). The height of the shelter is 2.83 m (9.2 ft.). The total weight of the shelter, including packaging for shipping, is 160 kg (352.7 lb). The shelter is designed to withstand winds of 28 m/s (101 km/h). It has passed the artificial rain test with no leakage. Moreover, it can withstand a minimum 2-min escape time in case of a fire hazard. The term "escape time" describes the amount of time available to (and required for) a person to escape a hazardous circumstance. In cold

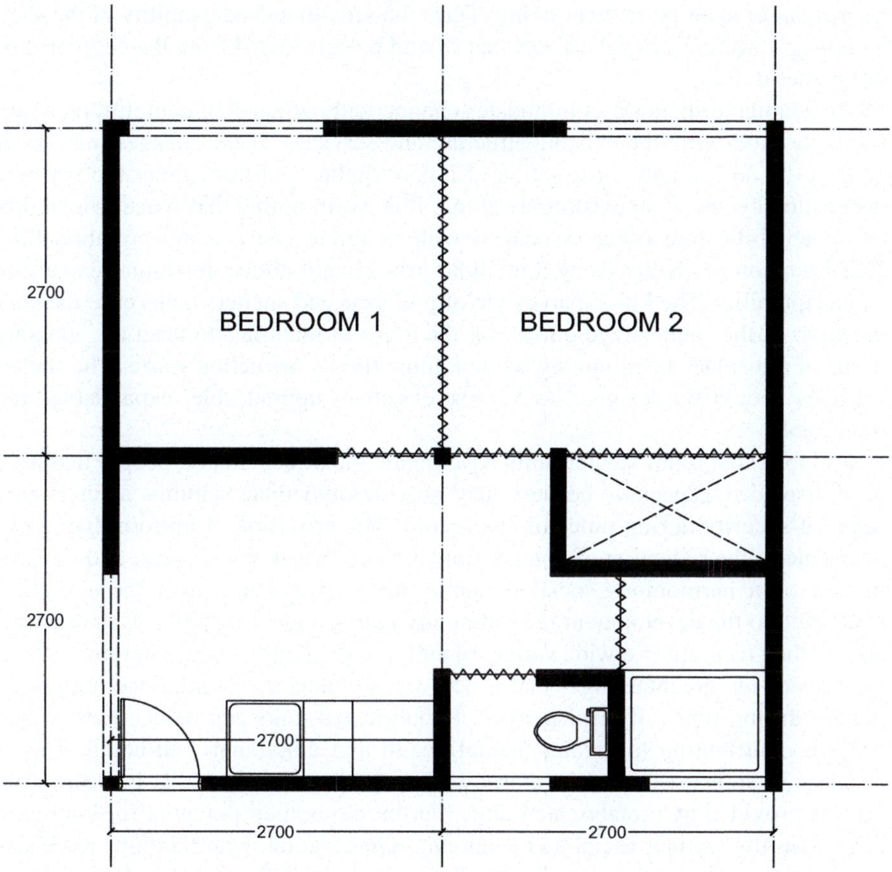

Fig. 7.6 Plan of prefabricated shelter

regions, it can withstand a 10 kg/m^2 snow load. The prefabricated shelter can be shipped to different countries via air, road, and sea from Poland, Panama, and Turkey. Forty-eight units can fit in a 40-foot-high cube shipping container, and a truck can ship 52 shelter units (BetterShelter n.d.).

7.3 Social Growth and Adaptability Using Prefab Elements

It is essential to understand the living problems of refugees. The first issue is the social and cultural problems that refugees face. Although sheltering is the initial requirement of the refugees, their homeless situation is unexpected and sudden, so they often seek immediate shelter to protect and keep their families living together. When the refugees start living in the host country for years after years, the shelters they started living in require adaptability to merge with their new demands for permanent or semi-permanent living. Thus, the growth and adaptability of the shelter designs are very crucial aspects that should be considered from the beginning of the shelter design.

An adaptable shelter accommodates change without needing demolition or substantially modifying the existing structure and services. These changes may occur both in the design and construction of a new shelter and throughout the multiple occupation cycles of an existing dwelling. It is worth noting that space's flexibility and adaptability are often associated with modular systems and prefabrication. Prefabrication of shelter using a modular grid system allows for future expansion and adaptability. The kit-of-parts style of prefabricated shelters helps ease the construction of the shelters and allows for the reuse of materials to meet the adapting needs of a displaced community while minimizing construction waste. The shelter modules should be designed to be cost-effective, upgradeable, expandable, and relocatable.

Prefabricated relief shelter units have major social benefits on people that have been displaced, especially because they provide immediate stability, safety, and a sense of security during times of emergency. The provision of uniform living circumstances, the reduction of competition and conflict over resources, and the promotion of a harmonious social dynamic are all ways in which these shelters contribute to the development of community cohesiveness. Because of their adaptability, they can satisfy a wide variety of cultural and family needs, thereby protecting privacy and maintaining social peace, all of which are essential for preserving dignity during times of displacement. Prefabricated units can reduce tension and anxiety, contributing to greater mental health and emotional well-being. This is because, in comparison to improvised shelters, an environment that is cleaner and safer is provided by prefabricated units. Furthermore, their potential for long-term usage permits gradual social and economic reintegration, which in turn facilitates smoother transitions from temporary housing alternatives to permanent housing options. In general, these units play an important part in both the immediate relief efforts and the longer-term recovery efforts, which helps to strengthen the social resilience of communities that have been evacuated.

7.4 Case Study: Prefab Housing for Tohoku Earthquake-Affected Communities in Japan

On March 11, 2011, a massive number of towns of various sizes in three prefectures—Iwate, Miyagi, and Fukushima—all in the Tohoku area, were impacted by the Great East Japan Earthquake, which measured 9 on the Richter scale and was the fourth largest earthquake to strike the planet since 1900. More than 15,000 people were killed, more than 129,000 homes were destroyed, and more than 266,000 were damaged. The Fukushima nuclear facility experienced a nuclear disaster that reached a radioactive contamination level of 7, forcing 90,000 persons in the districts near the stricken nuclear plant to evacuate. The earthquake created a tsunami that inundated and destroyed 56,100 hectares (JICA (Japan International Cooperation Agency) 2013).

All the prefectures had contracts with the JPA prefabrication businesses, which produced most of the 43,000 units. The Japan Federation of Home Production Organizations and the Japan Wooden Housing Industry Association, among other businesses, had to be hired due to the scale of the calamity (more than 50,000 temporary homes were produced). More than 6000 units were constructed by the later timber manufacturing firms, with 4000 near Fukushima. The prefab homes were constructed in around 2 months, and the timber homes took a little longer (Bris and Bendito 2019).

Prefab light metallic structures were employed to build most of the houses. Since 26 separate companies in Fukushima alone worked on creating these sorts of homes, there was a far lower percentage (12%) of houses made with wood, even though there were a lot of them (about 6000) (Haganuma et al. 2015). The building quality was very poor, and a variety of assembly issues, including gaps between walls and roofs, draughts, lack of shelves and storage areas, places to sit outside, lack of an enclosure, and lack of a veranda at the entrance door, occurred because of the large number of prefab houses that needed to be built, the use of minimum standards, and the tight deadlines for assembly. Residents were interviewed, and concerns about noise, condensation on windows, difficulty dispersing heat in the summer, inadequate insulation, and lack of ventilation under the floor (resulting in mildew) were all noted. Three types of shelters were designed with bedroom areas of 20 sq. m (200 sq. ft.), 30 sq. m (300 sq. ft.), and 40 sq. m. (400 sq. ft). The arrangement, which included the pattern for the wooden models, was relatively identical. Unusually, a bigger model (four bedrooms) was offered in the wooden versions for the larger families leaving Fukushima. All the types of shelters were designed modular and have provisions for future adaptability and expandability (Haganuma et al. 2015).

Due to the size and scope of the tragedy, each temporary housing site had a different amount of temporary homes. Each temporary housing site can store up to 250 homes in cities, unlike rural regions, which are smaller and can only hold 50–60 homes, like in Sendai, where eight temporary housing could accommodate 1505 homes. As a part of an upgrade in 2004, a community centre for every 50 temporary houses and, in some instances, additional installations were built to provide care for the most vulnerable people (Bris and Bendito 2019).

7.4.1 Lessons Learned

- **Prefabrication for emergency response**: Prefabricated shelters offer a fast, cost-effective solution during emergencies, as they are partially or entirely constructed in factories and then assembled on-site. This process saves time, ensures better quality control, and reduces material waste, making it ideal for disaster situations.

- **Community involvement through self-help kits**: Prefabricated self-help kits enable refugees to take part in the construction of their own homes, fostering a sense of ownership and dignity. This method reduces labour costs and engages the community in decision-making.
- **Versatility in shelter solutions**: Prefabricated shelters can serve multiple purposes, from emergency to permanent housing. Modular systems and flexible designs allow for future expansion, relocation, and adaptation, ensuring long-term usability and reducing environmental impact.
- **Plug and Play shelters**: Plug and Play prefabricated units offer an innovative solution for rapid deployment in emergencies. These modular units require minimal assembly, are easily transportable, and can be set up quickly, making them ideal for immediate use in disaster zones.
- **Container shelters for refugee communities**: Prefabricated container shelters are effective for large-scale housing needs in refugee camps. These shelters offer comfort, durability, and customization options and can be rapidly deployed in response to disasters.
- **Adaptability and growth**: Prefabricated shelters should be designed with adaptability in mind, allowing for future growth and modifications. This is particularly important for long-term refugee housing solutions, as the needs of the displaced community evolve over time.
- **Sustainability and waste reduction**: Prefabricated shelters generate less waste compared to traditional construction methods. Modular and kit-of-parts systems allow for the reuse of materials and components, reducing environmental impact and promoting sustainability in shelter design.

Weblink of Case Study Project: https://www.researchgate.net/publication/276519622_A_STUDY_ON_THE_REUSE_CHARACTERISTIC_OF_WOODEN_TEMPORARY_HOUSING_-_A_study_on_wooden_temporary_housing_in_fukushima_prefecture_after_the_tohoku_earthquake_-

7.5 Conclusion

Considering construction time and process, prefabricated dwelling units are often used as an effective solution for emergency shelters. Prefabricated prototypes, mostly ready-made structures, are shipped to the disaster area during the emergency, either in pieces or fully assembled and built on-site. Prefabrication can also ease the self-help construction process. The future of prefabrication and rapid construction will unfold with building information modelling software. Prefabrication methods such as modular, panelized, and kit-of-parts systems ensure high-quality construction, minimize waste, and reduce labour costs, while also allowing for flexibility in design and future expansion. Engaging refugees in the shelter-building process through self-help kits fosters a sense of ownership and community involvement, which is vital for their well-being. Although there are challenges, such as higher upfront costs, transportation logistics, and the need for climate-appropriate designs, prefabricated shelters can be tailored to meet cultural, social, and environmental considerations. By incorporating adaptability, sustainability, and scalability into their design, these shelters can provide not only immediate protection but also long-term, dignified living solutions for displaced populations.

Discussion Questions

- What are the main advantages of using prefabricated dwelling units for emergency shelters?
- How does the prefabrication process ease the self-help construction process for refugees?
- In what ways do prefabricate shelter kits offer benefits in terms of transportability and assembly?
- What are the key design factors to consider for prefabricated emergency shelters, such as modularity and adaptability?
- How can prefabricated shelters be expanded or modified to suit different environments and needs?
- What are some of the disadvantages of prefabricated shelters, and how might they impact their effectiveness?

References

BetterShelter (n.d.) Relief housing unit (RHU). https://bettershelter.org/rhu/. Accessed 10 Jul 2022

Bris P, Bendito F (2019) Impact of Japanese post-disaster temporary housing areas' (THAs) design on mental and social health. Int J Environ Res Public Health 16(23):4757. https://doi.org/10.3390/ijerph16234757

Chauhan K, Peltokorpi A, Lavikka R, Seppänen O (2022) Deciding between prefabrication and on-site construction: a choosing-by-advantage approach performance evaluation of construction companies and suppliers view project. https://doi.org/10.24928/2019/0158

Friedman A (2021) Pre-fab living. Thames & Hudson, London/New York, NY

Gupta M (2012) Settlement and shelter reconstruction. Handbook of hazards and disaster risk reduction :553–566. https://doi.org/10.4324/9780367854584-49/SETTLEMENT-SHELTER-RECONSTRUCTION-ERIN-CONNELL-BRENT-DOBERSTEIN

Haganuma S, Urabe T, Ishisaka K, Wood Studio Co H (2015) A study on the reuse characteristic of wooden temporary housing—a study on wooden temporary housing in Fukushima prefecture after the Tohoku earthquake. J Archit Plann 80(710):813–822

JICA (Japan International Cooperation Agency) (2013) The study of reconstruction processes from large-scale disasters. Tokyo, Japan

Karmod (n.d.) Refugee Housing Unit | Asylum-seekers and Migrants. In: Karmod. https://karmod.com/en/page/refugee-housing-unit. Accessed 10 Jul 2022

Keena N, Friedman A (2024) Sustainable housing in a circular economy. Routledge, UK

Lopez D, Froese TM (2016) Analysis of costs and benefits of panelized and modular prefabricated homes. Procedia Eng 145:1291–1297. https://doi.org/10.1016/J.PROENG.2016.04.166

Melis L (2003) Parasite paradise: a manifesto for temporary architecture and flexible urbanism. NAi Publishers/SKOR; Available in North America through D.A.P./Distributed Art Publishers, Rotterdam/New York, NY

Roberts T, Writer R (2020) Pros & Cons of prefab homes. In: Build with rise. https://www.buildwithrise.com/stories/pros-cons-prefab-homes. Accessed 10 Jul 2022

Shaw M (2015) These modular homes enable plug-and-play urban preservation. Architizer J. https://architizer.com/blog/inspiration/stories/pao-courtyard-house-plugin/. Accessed 15 Jul 2022

The Guardian (2017) Why Ikea's flatpack refugee shelter won design of the year | Design | The Guardian. https://www.theguardian.com/artanddesign/2017/jan/27/why-ikea-flatpack-refugee-shelter-won-design-of-the-year. Accessed 10 Jul 2022

UNHCR (2024) Emergency shelter solutions and standards. https://emergency.unhcr.org/emergency-assistance/shelter-camp-and-settlement/shelter-and-housing/emergency-shelter-solutions-and-standards. Accessed 2 Oct 2024

Chapter 8
Designing Shelters for Self-Help Building

Keywords Self-help refugee housing · Emergency shelters · Refugee housing · Disaster recovery shelter · Architect's role in humanitarian architecture · Sustainable refugee housing solutions · Self-help construction training · Shelter materials selection · Transitional shelters · Liina shelter case study

With the increasing incidence of natural calamities and civil wars, one of the vital concerns for architects and planners is the effort to establish a logical and immediate response to emergency shelters. As a rapid response during the initial emergency, the government or humanitarian organizations provide the materials and aid for shelter seekers, and displaced people use shelter kits to build houses independently. Therefore, in different circumstances, such community-led housing initiatives are widely studied topics in developing and developed countries. In the past, the role of architects for aided and humanitarian architecture was more focused on operational and technical concerns rather than design thinking. Strategic spatial problem-solving and architecture-driven approaches for long-term recovery in devastated communities were missing at that time. The necessity of architects addressing self-help housing challenges following a disaster or war has thus become an evolving issue. Architects' efforts in self-help housing allow refugees and people affected by natural disasters to build houses. However, with additional consideration of design ideas for emergency housing, their contribution can rebuild the local culture, environment, and economy of damaged communities and ensure they are longer-lasting and better performing.

This chapter outlines the concept of self-help housing, its various types (independent, organized, employed), and its benefits, including financial savings and empowerment of residents. It also delves into the history of self-help housing, its evolution, and its use in refugee housing since the 1920s. The potential of self-help housing as a mid-term solution for refugees is discussed, emphasizing the role of organizations like UNHCR. Guidelines for designing and constructing self-help shelters are provided, including structure type selection, construction material

A. Friedman, B. Chaki, *Fundamentals of Planning and Designing Sustainable Post-Disaster Shelters*, https://doi.org/10.1007/978-3-031-83317-5_8

choice, and specialized work processes. The chapter highlights the importance of construction training and quality assurance, along with a case study showcasing the Liina transitional shelter, a sustainable and adaptable self-help housing solution. Overall, this chapter underscores the significance of self-help housing in addressing the urgent need for refugee shelters while promoting self-sufficiency and community rebuilding.

8.1 What Is Self-Help Housing?

Self-help is a practice of housing construction that emerged long before architect John Turner articulated his theory (Ward 1983; Harms 1992; Parnell and Hart 1999). A revolutionary idea in the late 1950s in housing that changed the whole view on housing for the poor. Turner demonstrated in his fieldwork in Lima and Peru how people used their resources and ingenuity to build 50,000 residential units over 7 years (Ward 2012). The critical factor of this initiative was to fulfill their own housing needs, of which no societies are exempt (Ward 2012). Self-help housing continues to be successful in reducing costs, improving neighbourhoods, increasing family satisfaction, and facilitating bonding between occupants and their homes.

According to a definition given by the self-help organization, "Self-help housing involves groups of local people bringing back into use empty properties that are in limbo, awaiting decisions about their future use, or their redevelopment" (Self-Help Housing Organization n.d.). Self-help housing strategies also denote individuals renovating or repairing their own houses using their labour. People in low-income places employ self-help mostly because they have no other choice. However, in more developed countries, self-help is a strategic approach that offers certain advantages (i.e. lower construction cost as it requires no extra labour cost, building a strong relationship with the home, providing an opportunity to customize as the self-builder choice) and encourages moving in this direction (Ghandehari 2000) (Fig. 8.1).

8.2 Types of Self-Help

According to Ward (1983), there are three predominant types of self-help housing typologies depending on the coordination and organization of the self-help process (Ward 2012).

1. In independent self-help, individuals carry out various procedures without the assistance of external sponsorship, supervision, or government funding. Without being a part of a program or a group, dwellers of residences construct, manage, finance, design, and make all choices regarding their homes. Consequently, indi-

Fig. 8.1 Young adult participating in self-help construction

viduals contribute "sweat equity" through their work and "business equity" as their contractors (Rueda 2018).

2. In organized self-help and mutual help, low-income users are sponsored, monitored, or assisted by external agents who commence the self-help activities. The participants undergo a pre-construction orientation and training before building and occupying their completed homes (Rueda 2018).
3. In employed self-help, an organization launches a self-help program in which members are hired and paid cash compensation while building their homes. Government entities frequently support this to assist low-income families (Rueda 2018).

Regarding refugee shelters and housing, organized self-help, mutual help, and employed self-help are mostly followed. Most often, after a conflict or natural disaster, the refugees have nothing to finance and have to build houses using their own money. Therefore, for many decades, different shelter organizations, such as the Global Shelter Cluster and the International Organization for Migration (IOM), have sponsored and organized self-help shelter-building programs to provide humanitarian shelter and settlements to people affected by natural calamities and conflict. These organizations work on operations and coordination and play a significant role in supporting national authorities in collaboration with partners.

8.3 Benefits of Self-Help in Refugee Housing

Self-help construction has different advantages for housing refugees: financial, physical, and social advantages (Ward 2012). Although financial advantages are often the most alluring reason governments and organizations adopt self-help in refugee shelter construction, other advantages should not be ignored. The most significant benefit of self-help is that it saves the construction cost of housing using sweat equity and enterprise equity. Still, the refugees engaging in self-help construction can enjoy working independently according to their budget and time.

Self-help contributes to additional housing services to address homelessness and meet the needs of emergency housing and mid-term transitional housing. It helps to meet the shelter needs refugees' shelter needs when they have nowhere to live in the host community. Self-helpers can acquire construction skills and qualifications through on-site experience. It offers the benefits of active participation, increased self-confidence, and control over crucial aspects of their lives.

Shelter organizations such as UNHCR can help to address empty properties and build harmony within the community. Addressing negligence and integration through "street-level programs" can lead to broader community regeneration that engages people through wider policy goals.

8.4 History of Self-Help

The idea of self-built housing for low-income groups of people developed in stages. The Swedish government implemented this policy in 1904, and various programs were arranged throughout Europe and the Soviet Union in the 1920s (Harris 1999). Such interventions and programs were proactive responses to specific historical situations, and most of them were for a short duration. The only lasting plan was the one launched in 1926 by the City of Stockholm (Harris 1999). It did not appear in this era that any of those who supported such programs had any broader idea of the role of aided self-help within a continuing housing policy. It also remained the same in the 1930s, when many governments, including Europe and North America, introduced self-help into homestead schemes due to the great depression (Harris 1999). More consistent policy and the associated theory of aided self-help emerged during the 1940s. The first significant initiative, an aided self-help housing project, was completed in Puerto Rico by the U.S. government. Since then, aided self-help has been practiced in developing and developed nations as a successful way to provide housing for the homeless, refugees, and low-income groups (Harris 1999).

Self-help schemes have been used for refugee housing since the 1920s. In 1922–1923, Greece began receiving thousands of refugees following its defeat by Turkey in Asia Minor. A Refugee Settlement Commission, formed by the Greek government under the guidance of the League of Nations, implemented several policies to address the housing crisis (Kapur 1989). One of the policies included offering individual family sites, basic public infrastructure, and land and mortgage loans to self-help cooperatives aided by self-help strategies. Such a pragmatic approach was directed at refugees when tens of thousands of Greek families constructed their homes. Although some national and local governments were more hesitant in their approval, the self-help approach proved to be a successful strategy for refugee housing (Pooley and European Science Foundation 1992). Another organization, the Tamil Community Housing Association, was founded in response to the mass refugee movement that followed immediately after the refugee crisis in Sri Lanka during the early 1980s (Teasdale et al. 2009). Eventually, the association focused on the settled families for developing their houses permanently and expanding their goal to make the refugee community an asset (Teasdale et al. 2009). Although the Tamil Community Housing Group has shifted its focus from refugee settlement, it still provides housing in empty land using self-help strategies (Teasdale et al. 2009).

Aided self-help housing has two main advantages, making it a beneficial and popular housing for displaced persons. First, it has the potential to spread limited funding as widely as possible, making the best use of it for large-scale investments, where the individuals are free to do what they can to improve their houses. This makes aided self-help housing appealing to governments and NGOs such as the World Bank and other humanitarian organizations (Mullins 2018). Moreover, it also has significant advantages for governments of underdeveloped and developing countries by not requiring severe and unsustainable financial commitments to fund housing for the poor (Mullins and Moore 2018). By endorsing self-help housing, governments and organizations in developing countries can save considerable public funds by supplying serviced packages and leaving the rest to be finished by the residents themselves (Nakamura 2014). Moreover, experts and the state in developing countries have embraced Turner's ideas and initiatives on aided self-help housing primarily because they seem practical and realistic and include the prospect of large-scale savings in public housing budgets. In another study, Mukhija (2004) noted that aided self-help housing is not only a cost and time-effective process but also encourages the donors to sponsor affordable housing for refugees and low-income people as it requires zero to low labour costs (Mukhija 2004). Despite these benefits, aided self-help housing is not beyond criticism. Several authors indicated that self-help by site and service schemes was generally ineffective as a low-income housing strategy (Sundaram 2012; Bredenoord 2021).

8.5 The Potential of Self-Help Refugee Housing as Mid-Term Solution

Despite ranging from informal community housing projects to social enterprises, self-help housing strategies can be a holistic approach to housing and employment provision for refugees by incorporating their labour and skills without requiring professional training (Teasdale et al. 2009). After providing immediate shelter for the refugees, they are transferred to transitional houses with the expectation of staying there for long term. These houses provide amenities and facilities as a mid-term housing solution until the refugees return to their country of origin or resettle. Housing by self-help is popular for providing mid-term housing solutions in the host country.

The United Nations High Commissioner for Refugees (UNHCR) has worked for refugees since the 1950s. During a crisis, they provide adequate shelter to refugees. They incorporate self-help strategies and involve refugees in the building process of shelters by providing the building materials, tools, and guidelines for their construction. Generally, they distribute emergency shelter kits from three of their centres in Dubai, Copenhagen, and Durban. In the primary phase, they provide shelter kits with tents, plastic sheeting, and matting so that the refugees can build very quickly and easily, and thereby, they do not need to spend a night under the sky. In longer-term circumstances, they fund the renovation of community shelters or the production of new homes and provide refugees with the resources they need to build their own homes under self-help schemes. Additionally, they monitor the distribution and construction of the houses and manage water supply from surface water, rainwater, groundwater, natural reservoirs, or municipal or private systems if allowed (United Nations High Commissioner for Refugees 2007). For human excreta disposal, they arrange different systems according to their need. A well-laid-out camp protects the environment and helps to prevent fires and disease outbreaks. Although UNHCR arranges houses for emergency needs, many disease outbreaks appear when refugees start living in the host countries for several years. Therefore, there is a severe need to design and build the camp not only for an immediate response but also as a preventive measure for the future to ensure that their homes can be safe in case of long-term stays. As self-help has been proven to be a practical scheme for sheltering refugees, a well-designed house built by the people can also hinder other problems of emergency housing.

8.6 Guidelines for Self-Help Shelter

A self-help building process by low-income people following an emergency crisis appears very commonly in developing countries. It was used in various emergencies (natural disasters, war, or persecution) as an immediate, fast, and inexpensive solution worldwide. While designing self-help shelters for refugee housing, designers need to consider many factors ranging from the longevity of the shelters, construction time, and construction types. Below are the guidelines for designing and constructing a self-help shelter (Chaki and Friedman 2021).

8.6.1 *Structure Type Selection*

After a crisis, it is essential to build the structure within an immediate period, while making it usable in the long term. Considering these criteria, a prefabricated shelter structure can be a good option for refugee shelters. Furthermore, prefabrication makes the self-help construction process easier and minimizes material waste. Additionally, in cases where shelter kits are needed to be transported by UNHCR or other humanitarian organizations due to the emergency needs in a conflict or disaster zone, prefabricated shelter kits could be a viable option because of their easy transportability and convenient assembly process, all accomplished within less time than other kinds of shelters (Chaki and Friedman 2021). In the case study of Liina Transitional Shelter (Sect. 8.8), the modularity of the design allowed for efficient distribution of the shelters to crisis-affected areas.

For the long-term stay of refugees in host countries, the adaptability of the shelters is a critical issue. Hence, modular shelter design can be used as it allows future adaptability and minimizes material waste. Also, for prefabricated shelter construction, following modularity in the shelter design is highly recommended. Figure 8.2 shows how a modular grid pattern can be utilized to design the interior spaces of the shelters.

8.6.2 *Selection of Construction Material*

For any refugee shelter, the primary criteria for selecting materials for the structure and exterior envelope are durability, construction time, availability, and material cost. Also, for self-help construction, the material should be compatible with the refugees to avoid unwanted injuries. After the conflict, the shelter kits are transported to the conflict area by the UNHCR or other agencies. Therefore, materials that are easily transportable in shipping containers should be used. Considering

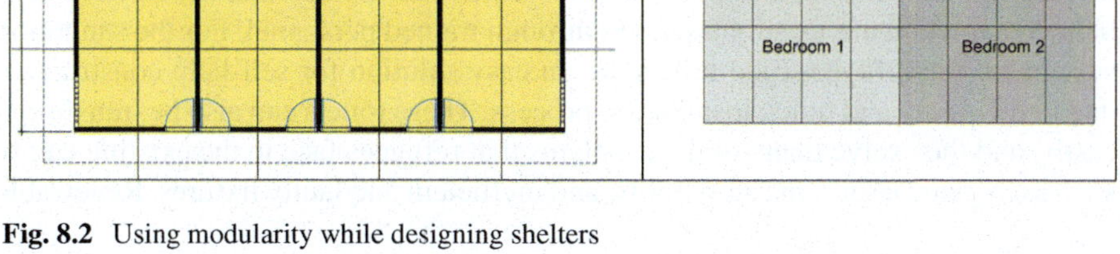

Fig. 8.2 Using modularity while designing shelters

these criteria, wood or timber enables transport without any hassle while ensuring durability and protection for a longer period. However, checking which materials are available in the surrounding markets for self-help construction is necessary. Additionally, prefabricating the shelter materials according to design can ease the self-help construction process and reduce construction time (Global Shelter Clusters 2018).

While selecting interior materials for self-help construction, such as walls, floors, doors, and windows, attention should be given to durability, protection from an adverse climate, construction time, ease of handling the interior component, and availability. Also, the materials should be prefabricated in a workshop for ease of construction. Generally, bamboo-framed doors and windows are common in many developing countries (i.e. Bangladesh and Myanmar) for self-help refugee shelters. Moreover, if the shelter kits are transported by the organization, prefabricated wood floors, timber-framed doors, and windows are sustainable materials for the immediate construction, ensuring the permanence of the structure (Chaki and Friedman 2021).

8.6.3 Erection of Building On-Site

The shelters' components are transported to the construction site in shelter kit packages by a humanitarian organization and distributed among the beneficiaries. Each piece in the shelter kit package should be numbered and indexed to step-by-step manuals understandable for any shelter owner without knowledge of architectural plans or the construction process (Fig. 8.3). Furthermore, the shelter organization should provide self-builders with help and instructional videos during construction. The designers and manufacturers of self-help shelters need to assess the shelter design to see if the packages can be assembled safely and without too much complexity (Rueda 2018). Moreover, to ensure a safe self-help construction process, the shelter organization should define the level of complexity with the shelter construction guidelines and check with the beneficiary if they can accomplish the construction (Rueda 2018).

8.6.4 Specialized Work Process

Specialized services such as plumbing and electrical and mechanical work in self-help construction are usually provided through trained personnel. For the sanitation system for the refugees, pod toilets are an easy solution for self-help construction due to their easy and quick installation process. These toilets can also be transferred easily, and they solve many of the problems that refugees face in the existing camp, such as hygiene issues, inaccessibility, and inefficient and faulty fixtures. Renewable

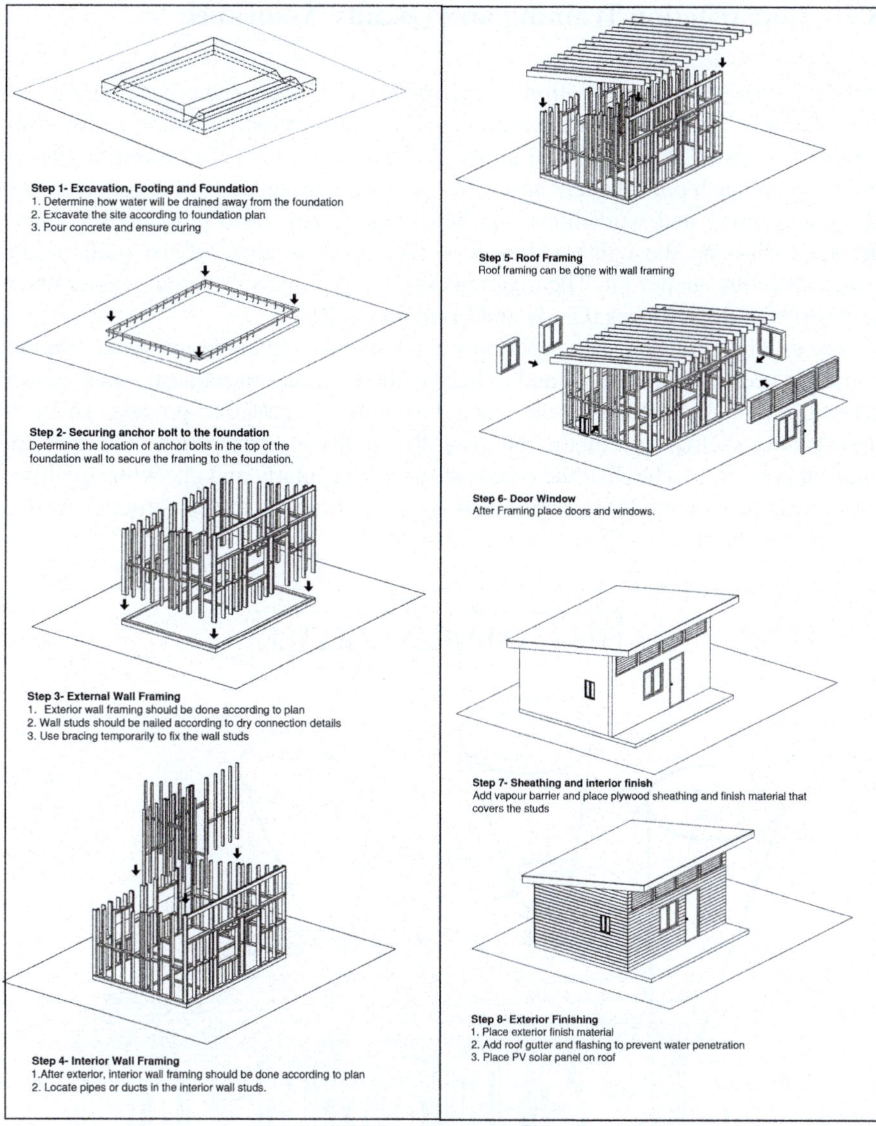

Fig. 8.3 Step-by-step diagram showing the assembly process

energy sources are very sustainable sources of energy for the refugee camps. Therefore, the design of the shelters and master plan of the site should include a solar panel on top of the roof or in any other suitable location in the camp. However, the electricity lines should be connected by trained electricians in each shelter as they require specialized training (Chaki and Friedman 2021).

8.7 Construction Training and Quality Assurance

Before the self-help construction of the shelter commences, the beneficiaries first need training on the construction process. In many cases, including in Rohingya refugee camps in Bangladesh, training of the trainers (ToTs) is given so that the trainers can teach other beneficiaries how to construct and assemble the structures. Besides training and workshops, the beneficiaries also need guidebooks or infographics showing the construction processes and assembly of the prefabricated materials in the shelter kit. The diagrams make the process easier and better understandable for the refugees (Chaki and Friedman 2021).

The organization and shelter partners need to make regular home visits to beneficiaries to keep them informed and to help with any issues or requests, such as work schedules and any difficulties they face during the construction process. To ensure high-quality shelter built by the refugees, the shelter organization should have specialized personnel who visit the construction site regularly and check the quality of the construction work. If any significant faults are found in the construction, trained workers are involved in fixing the issue (Fig. 8.4).

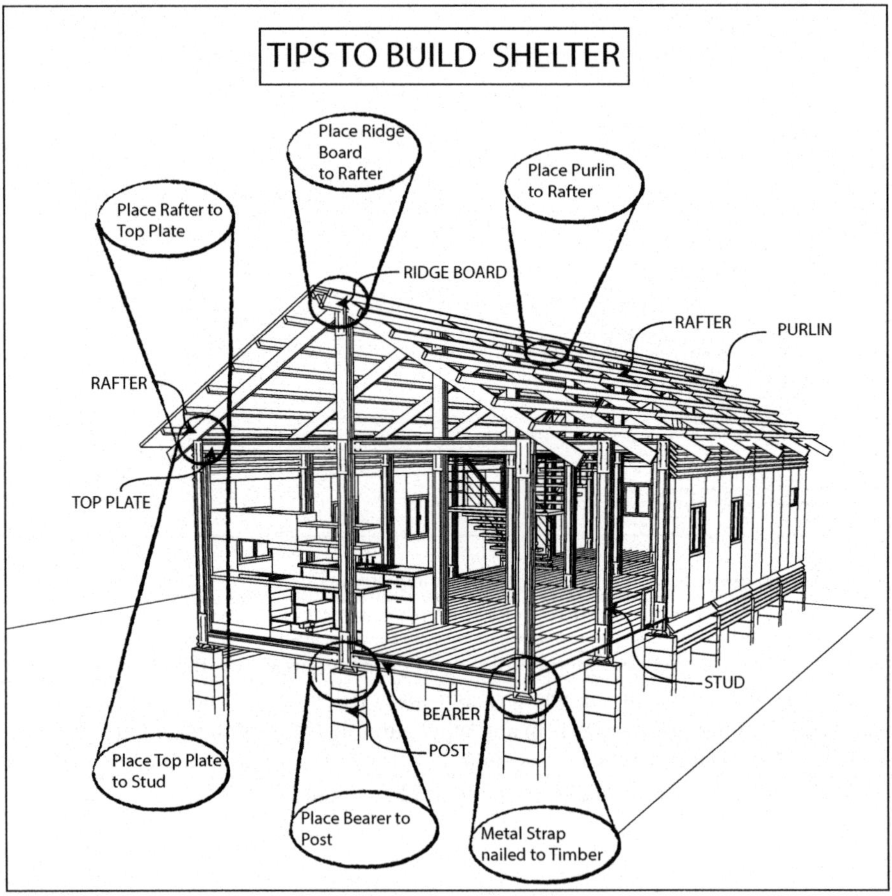

Fig. 8.4 Diagram showing basic construction technique for wood shelter

8.8 Case Study: Wooden Prefabricated Temporary Housing, Aalto University Wood Program

By 2050, the global refugee population is projected to exceed one billion. The Liina transitional shelter is a temporary structure designed for usage in cold regions and emergency scenarios worldwide. Designed and constructed entirely by students in the Wood Program at Aalto University, Liina is composed primarily of Finnish wood and wood-based materials, making it not only an elegant and dignified but also a sustainable solution to this worldwide issue (ArchDaily n.d.).

As a transitional type of self-help shelter, it provides a secondary response following a crisis. Two adults can construct the shelter in 6 h using only basic equipment and an instructional diagram. It is designed to accommodate a family of five for up to 5 years during the reconstruction period following a natural disaster, after which it can be renovated, relocated, resold, or recycled. The interior space is 18 sq. m, meant to meet the established requirement of 3.5 sq. m. (12 sq. ft.) per refugee. The inside features two semi-private sleeping areas, a small kitchen, and a dual-functioning dining/living/work area. A loft above the bedroom can be utilized as extra sleeping and storage space. An additional 7 sq. m. of enclosed outdoor space intended to serve many purposes for the occupants during their tenure (ArchDaily n.d.) (Fig. 8.5).

The Liina concept is built on a series of readily transportable and assemblable sandwich panels that may be flat-packed into a tiny space. Each panel is 600 mm (23.62 in.) wide (half the width of a sheet of plywood) to reduce waste and make it easy for one adult to transport. Six panels are attached with simple, repeating wooden joints to form a frame, then tightened with nylon straps ("liina" in Finnish), similar to those used in cargo shipping. The strap establishes airtight connections without using metal fasteners or electric tools on-site. The frame is then turned up,

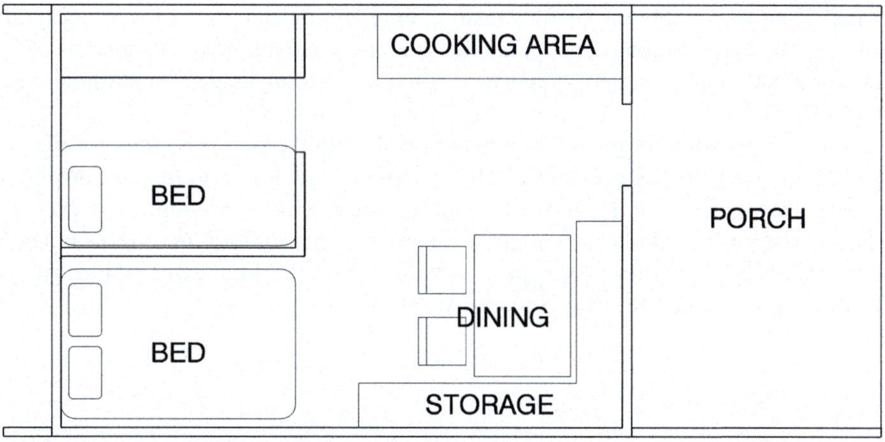

Fig. 8.5 Floor plan showing self-built Liina transitional shelter

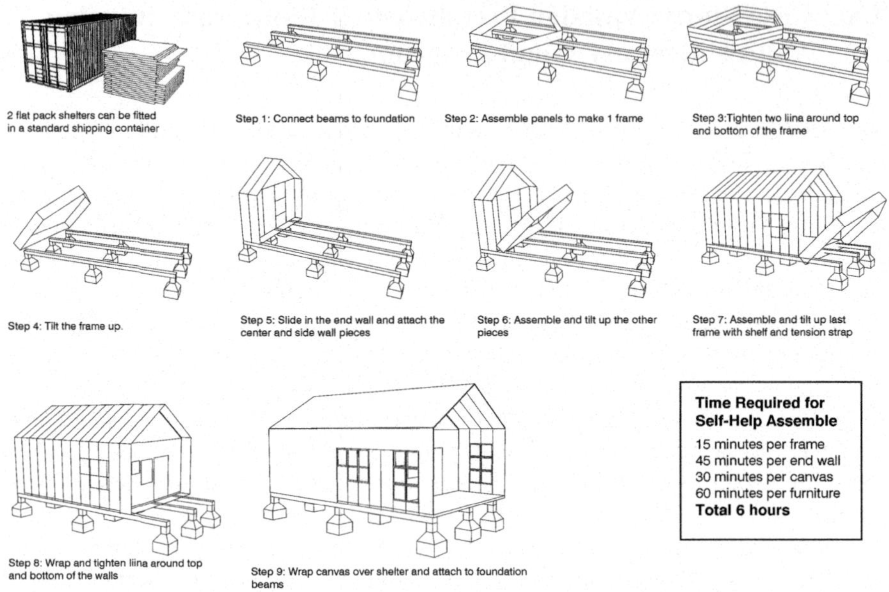

Fig. 8.6 Step-by-step assembly diagram for self-built Liina transitional shelter

stacked one on the other, and secured with three straps. A tent-like canopy protects the structure against water and ultraviolet harm. The modularity of the design permits the addition or subtraction of frames to increase or decrease the shelter's size to meet the inhabitants' needs. The shelter's design is intended to provide diverse solitude and areas for daytime and evening activities (Fig. 8.6).

Other design considerations include the panels' size and weight, their assembly simplicity, and their heat-insulating properties. The outcome is a low-tech solution consisting of plywood and LVL (laminated veneer lumber) panels insulated with wood fibre. In collaboration with a team of engineering students, 13 wood program students from ten nations constructed the shelter by making full-scale models of the building and conducting structural tests, such as earthquake and wind tunnel tests (ArchDaily n.d.).

Liina is intended for the environment prevalent in the Ararat region of Turkey, a location chosen for its continental climate and history of refugees and displaced people owing to natural calamities such as earthquakes, drought, and political unrest. However, Liina's design easily adapts to other cultural and environmental circumstances. A shelter prototype was built during the May 2012 World Village Festival in Helsinki's Railway Square (ArchDaily n.d.).

8.8.1 Lessons Learned

- **Cost-effectiveness**: Self-help housing significantly reduces the cost of construction, as labour costs are minimized or eliminated entirely. With materials provided by humanitarian organizations, refugees can use their labour and local resources, making it a cost-effective solution for large-scale housing needs. This approach allows funds to stretch further, enabling more shelters to be built within a given budget.
- **Adaptability and scalability**: Self-help housing models, especially those that utilize prefabrication and modular design, offer flexibility for future expansions, modifications, or relocations. This adaptability ensures that shelters can meet the evolving needs of residents and be scaled as populations grow or as the nature of their housing needs change from emergency to long-term solutions.
- **Community and cultural sensitivity**: The success of self-help housing depends on the incorporation of cultural and social norms in the design and construction process. Understanding the specific needs of refugee communities, including privacy, security, and the role of shared spaces, ensures that shelters are not only functional but also culturally appropriate and sensitive to the diverse backgrounds of the inhabitants.
- **Collaboration and support from humanitarian organizations**: Successful self-help housing projects require strong partnerships between local communities, governments, and international organizations. Humanitarian organizations play a key role in providing materials, technical expertise, and logistical support, as well as ensuring that shelter designs meet global standards for safety and durability.
- **Holistic approach to community building**: Shelter is just one aspect of recovery; self-help housing projects must also consider the broader community context. Providing access to utilities, communal facilities, education, healthcare, and spaces for social interaction is critical to fostering a sense of community and improving the overall quality of life for refugees.

Weblink of Case Study Project: https://www.archdaily.com/174909/liina-transitional-shelter-aalto-university-wood-program

8.9 Conclusion

Self-help housing presents an effective approach to addressing the shelter needs of displaced populations following natural disasters or conflicts. By involving refugees in the construction of their own shelters, self-help housing not only empowers individuals and communities but also reduces costs, promotes self-sufficiency, and provides an adaptable solution for long-term habitation. The evolution of self-help housing, from its early stages to its current role in humanitarian efforts, demonstrates its potential to meet both immediate and mid-term housing needs, particularly in refugee contexts.

Architects and planners now play a crucial role in enhancing the design of self-help shelters, ensuring they are sustainable, resilient, and capable of adapting to cultural and environmental challenges. Prefabrication and modular design further simplify the construction process, making it accessible even to those with limited building experience. With proper training, guidelines, and support from humanitarian organizations like UNHCR, self-help housing can be a transformative tool in disaster recovery, offering not only shelter but also hope and dignity to displaced individuals. Ultimately, the self-help model serves as a sustainable and scalable approach to rebuilding communities, fostering resilience, and enabling long-term recovery.

Discussion Questions

- What are the key benefits of self-help housing for refugees and displaced people?
- How does self-help housing encourage a sense of ownership and self-reliance among refugees?
- What role do architects play in the design and construction of self-help emergency shelters?
- How can training and infographics aid refugees in the construction process of their shelters?
- What are the potential long-term impacts of self-help housing on local culture, environment, and economy?
- What are some challenges that might arise in self-help housing projects, and how can they be addressed?

References

ArchDaily (n.d.) Liina Transitional Shelter/Aalto University Wood Program. https://www.archdaily.com/174909/liina-transitional-shelter-aalto-university-wood-program. Accessed 25 May 2022

Bredenoord J (2021) Self-managed cooperative housing by mutual-assistance as introduced in Central America between 2004 and 2016 the attractiveness of the 'FUCVAM' model of uruguay. J Archit Eng Technol 6:188. https://doi.org/10.4172/2168-9717.1000188

Chaki B, Friedman A (2021) Housing the Rohingya refugees in Bangladesh: learning from the experiences of self-help emergency shelters cases. J Archit Des Construct Technol 2(1):1–10

Ghandehari H (2000) Kit homes as a self-help affordable strategy. Unpublished M. Arch Report, McGill University, Montreal, Canada

Global Shelter Clusters (2018) Bangladesh 2017–2018/rohingya crisis, chrome-extension://efaidnbmnnnibpcajpcglclefindmkaj/ https://shelterprojects.org/shelterprojects2017-2018/SP17-18_A13-A15-Bangladesh-2017-2018.pdf. Accessed 30 May 2024

Harms H (1992) Self-help housing in developed and third world countries. In: Housing policy in the socialist third worldpolicy in the socialist third world. Mansell, London

Harris R (1999) Slipping through the cracks: the origins of aided self-help housing, 1918-53. Hous Stud 14(3):281–309. https://doi.org/10.1080/02673039982803

Kapur P (1989) From ideas to practice: "Self-help" in housing from interpretation to application. Massachusetts Institute of Technology (MIT), Boston, MA

Mukhija V (2004) The contradictions in enabling private developers of affordable housing: a cautionary case from Ahmedabad, India. Urban Studies. https://doi.org/10.1080/0042098042000268438

Mullins D (2018) Achieving policy recognition for community-based housing solutions: the case of self-help housing in England. Int J Hous Policy 18(1):143–155. https://doi.org/10.1080/19491247.2017.1384692

Mullins D, Moore T (2018) Self-organised and civil society participation in housing provision. Int J Hous Policy 18(1):1–14. https://doi.org/10.1080/19491247.2018.1422320

Nakamura S (2014) Impact of slum formalization on self-help housing construction: a case of slum notification in India. Urban Stud 51(16):3420–3444. https://doi.org/10.1177/0042098013519139

Parnell S, Hart D (1999) Self-help housing as a flexible instrument of state control in 20th-century South Africa. Hous Stud 14(3):doi 367-386

Pooley CG, European Science Foundation (1992) Housing strategies in Europe, 1880–1930. Leicester University Press, Leicester

Rueda MC (2018) Prefabricated self-help kit homes- a mass customized option for affordable housing. Unpublished M. Arch Report, McGill University, Montreal, Canada

Self-Help Housing Organization (n.d.) What is self help housing? | Self Help Housing. https://self-help-housing.org/what-is-self-help-housing/. Accessed 22 May 2022

Sundaram A (2012) Impact of Self-help Group in Socio-economic development of India. IOSR J Hum Soc Sci 5(1):20–27

Teasdale S, Jones PA, Mullins D (2009) Briefing paper 53 connecting the dots: the potential for self-help housing to address homelessness. Management (January): pp 1–27

United Nations High Commissioner for Refugees (2007) Handbook for emergencies. United Nations High Commissioner for Refugees, Geneva

Ward P (2012) Self-help housing ideas and practice in the Americas. MIT Press, Boston, MA, pp 283–310

Ward PM (1983) Self-help housing: a critique. Alexandrine Press, Mansell Publishing Ltd., Marcham, Oxon

Chapter 9
Sustainable Materials for Shelters

Keywords Sustainable shelter materials · Refugee shelter design · Climate change · Low-carbon shelters · Ecological footprint · Building materials · Environmental impact · Bamboo construction · Green materials · Shigeru ban

Climate change can no longer be ignored with the constant rise in the average global temperature. In refugee camps, thousands of shelters built within a short time are more prone to contribute to global climate change if sustainability is not considered from the beginning of the camp planning and shelter design phase. To reverse the phenomena, designers are introducing shelter designs that reduce the use of materials with high-carbon content and release smaller amounts of greenhouse gases into the atmosphere. Such design is known to have a small ecological footprint, which can be defined as the metric allowing for the calculation of an individual's, a community's, or a country's yearly consumption of nature's resources, expressed in equivalent numbers of "planets." Building materials for shelters need to be selected carefully since a poor choice may adversely affect their surroundings over their lifetime. Their impact on the environment must be considered, from their harvesting to dismantling the building.

This chapter examines various building materials, illustrates how to evaluate their sustainability, and discusses conventional and some of the most environmentally responsible materials available. The chapter begins by highlighting the alarming rates of natural resource consumption and their detrimental environmental effects, such as habitat destruction and carbon emissions. It emphasizes the need to manage resource consumption through tools like the ecological footprint to safeguard ecosystems. The chapter explains that building materials must be selected carefully, considering their embodied energy values. It discusses various materials commonly used in refugee shelter construction, including natural timber, bamboo, adobe bricks, clay or burnt bricks, wattle and daub reinforced mud walls, corrugated galvanized iron, plastic sheets, thatching materials, cement, cement blocks, and quarried materials. Each material is assessed in terms of its environmental impact, highlighting the need for responsible sourcing, manufacturing, and disposal practices. Finally, the chapter provides a case study of Shigeru Ban's innovative Paper

A. Friedman, B. Chaki, *Fundamentals of Planning and Designing Sustainable Post-Disaster Shelters*, https://doi.org/10.1007/978-3-031-83317-5_9

Log Houses, which exemplify cost-effective, recyclable, and easy-to-assemble shelters deployed to aid disaster survivors worldwide.

9.1 Low-Carbon Shelters

Current rates of natural resource consumption have created an environment threatening the earth's biodiversity. Examples include collapsing fisheries, diminishing forest cover, depletion of freshwater systems, and the buildup of carbon dioxide emissions (Tu et al. 2019; Hussain et al. 2020). By using tools like the ecological footprint, people will be able to ultimately manage the consumption of resources in neighbourhoods better to ensure the survival of the world's ecosystems. At present, the carbon footprint represents 60% of humanity's overall ecological footprint, and this percentage continues to grow (Mancini et al. 2016). Therefore, the burning of fossil fuels is mainly responsible for global warming. Initiatives to minimize the consumption of primary energy and fossil fuels will effectively reduce the size of the carbon footprint and, by extension, the ecological footprint.

Building materials should be carefully chosen to build a low-carbon shelter for refugee housing. Certain materials have lower embodied energy than others. For example, collecting and manufacturing wood take much less energy than producing concrete. However, concrete provides more thermal mass than wood, which can reduce operating energy. Materials must, therefore, be cross-compared when designing a low-carbon shelter for their embodied energy values. Salvaged, recycled, and local materials may also reduce embodied energy in dwelling design (Dabaieh et al. 2020).

With global warming, the need to rethink the current consumption and production methods of natural resources is ever more apparent. Since energy-inefficient housing is a large contributor to greenhouse gas emissions, unbuilt refugee camps present an opportunity to plan ways to reduce the carbon footprint in their construction and maintenance. Low-carbon shelters will form sustainable neighbourhoods. Combining initiatives for greener dwellings with effective camp design policies will help conserve the environment for the long term (Dabaieh et al. 2020).

9.2 Past and Current Practices in Material Choices for Shelters

Shelters should be built with materials that must provide weather protection, living space, storage, privacy, and emotional stability. As needed, blankets, mats, and tarpaulin must be given. Wherever possible, refugee shelters should be culturally and socially suitable and familiar. If appropriate local materials are available, they should be used. The raw material extraction stage, production phase, transportation, installation or use, maintenance phase, and end-of-life stage are all interconnected processes in the life cycle of materials and, ultimately, a shelter. Below are some common materials used for refugee shelter construction (UNHCR 2021).

9.2.1 Natural Forest and Plantation Timber

Forest damage, soil erosion, landslides, land degradation, habitat destruction, and increased flood risk are all consequences of unsustainable or improper forest timber extraction. Timber production can be integrated with biodiversity conservation and ecosystem services like landslide avoidance and water supplies. Monoculture timber plantations can lead to land degradation, habitat and biodiversity loss, and hydrological issues in the case of timber plantations. Forests and rural roads can be harmed by transporting wood/logs. Therefore, it is crucial to source wood from forests that have been logged responsibly and in accordance with local forest management plans (McEwan et al. 2020).

Timber mills are places where wood is processed. Mills can pollute the environment with solid waste, inefficient timber use, and noise and air pollution. Many types of wood require treatment to keep pests at bay and moisture out. Toxic chemicals are used for treatment, posing environmental and health risks. Correct structural design for timber constructions should be established to avoid unnecessary waste and inefficient use of lumber, and timber quantity and quality should be calculated based on needs. It is necessary to make use of a timber profile that reduces off-cuts. Lumber sensitive to insect attacks should be treated with sustainable or natural surface treatments to maintain long-term durability. To reduce damage and waste, timber should be stored appropriately. Timber reuse and recycling from waste (e.g. door and window frames and roof components) should be promoted (Woodard and Milner 2016).

9.2.2 Bamboo

Natural or cultivated bamboo groves can be used to harvest bamboo. Bamboo harvesting without regulation may result in habitat loss and riverbank erosion. Bamboo is a viable alternative to lumber since it grows quickly, does not require replanting, and is renewable. Bamboo should be acquired from well-managed sources that do not harm other natural regions (for example, bamboo plantations should not be used to replace natural forests). Bamboo may help local livelihoods and preserve traditional knowledge relating to land-use practices (van der Lugt et al. 2006) (Fig. 9.1).

Bamboo may be utilized without any further processing, but if the intention is to make it last for a long time, bamboo should be treated. It may also be turned into high-quality goods, but some may need high-energy manufacturing techniques, mainly if the bamboo must be treated for long-term usage. Chemical treatments used in bamboo treatment can pollute the air and water and create health problems. Therefore, it is necessary to use non-toxic and approved chemicals to treat bamboo. Bamboo should only be utilized where harvesting is possible in a sustainable manner. Borax or boric acid treatment is widespread, but it should be done with caution and training because both compounds have known health risks. Bamboo from local

Fig. 9.1 Bamboo house

industry should be used to help local communities. To minimize damage and waste, bamboo should be kept carefully. Bamboo reuse and debris recycling should be encouraged (Zea Escamilla and Habert 2014).

9.2.3 Adobe Bricks

Adobe bricks need a lot of clayey soil. Soil extraction has the potential to destroy natural habitats and farms, contaminate water bodies, produce ponds where disease vectors can thrive, change the local hydrological regime, and induce soil erosion. Because the bricks are sun-dried, no pollution is created during the production/installation. Earth excavation and manufacture occur on-site or close to the building site, reducing the need for transportation. The production facility needs a large amount of room for drying and excavating. Brick manufacturing is a high-water-consumption business that needs a long-term, comprehensive water management approach to lessen dependency on freshwater (Fig. 9.2).

Fig. 9.2 Clay brick house

Adobe brick constructions are not acceptable in wet environments without sufficient rain protection and flood mitigation methods in flood-prone locations (Fig. 9.3). Filling and stabilizing material for roadways, and aggregates for concrete or mortar are some of the applications for down-cycled clay building materials (Salih et al. 2020).

9.2.4 Clay or Burnt Bricks

The manufacturing of burnt bricks necessitates a significant volume of clay. As with soil extraction for adobe bricks, clay mining has the potential to destroy natural habitats and farms, contaminate water bodies, produce ponds where disease vectors can thrive, disrupt the local hydrological regime, and induce soil erosion. In most sectors, burned bricks are produced in wood or coal-fired kilns. Kilns pollute the air because they burn coal and wood, generating a lot of CO_2. Forests are threatened by the need for firewood, which can lead to deforestation and erosion. Transportation may also contribute to CO_2 emissions and air pollution. Improved kiln technology should be encouraged to reduce pollution during brick manufacture (Riaz et al. 2019).

Standard lengths and suitable wall thicknesses should be utilized to reduce pollution during the building process. To prevent losses, good loading techniques during transportation should be adopted. Reuse or down-cycling of bricks should be promoted when a brick edifice is decommissioned (Joglekar et al. 2018).

Fig. 9.3 Adobe house with thatched roof

9.2.5 Wattle and Daub Reinforced Mud Wall

This walling technique requires a considerable amount of soil. Soil extraction can result in habitat damage, landslides, erosion, water body contamination, and hydrological changes. It also produces disease vector breeding ponds and extraction holes. A wooden or bamboo framework structure is required to build walls, putting strain on timber or bamboo supplies and perhaps causing uncontrolled growth.

Because the walls are constructed on the spot, transportation requirements are decreased. This walling solution does not pollute the air and does not necessitate using energy or mined materials; nevertheless, it does necessitate using water. The technique should only be used in areas where soil can be extracted without causing hazards or environmental impacts, and it should be accompanied by strategies for reestablishing the extraction area, preventing accidents, and controlling water accumulation (vector control) in the excavation areas. A decommissioned mud wall can create plant-growing surfaces (Chettri et al. 2021).

9.2.6 *Corrugated Galvanized Iron*

Corrugated Galvanized Iron (CGI) sheets are cold-rolled hot-dip galvanized mild steel with a linear corrugated pattern. Steel, zinc, and other metals are used in vast amounts in manufacturing processes, and mining has social and environmental consequences. CGI sheets are protected from corrosion by a zinc coating (or galvanized) applied to the steel sheet, and they are made using energy and water-demanding procedures, commonly with carbon emissions from fossil fuel energy sources.

Manufacturers pollute the air and water by releasing hazardous heavy metals into the environment. Steel manufacturers use a lot of water, and the majority of the water is lost due to evaporation during cooling. In corrosive settings under humid conditions, CGI sheets should be avoided. To minimize excessive waste, the number of sheets should be adequately measured. This is an excellent idea to encourage the reuse of uncorroded sheets from historic structures. To minimize damage, the sheets should be kept dry and stacked at appropriate heights (UNHCR 2021).

9.2.7 *Plastic Sheets*

Polyethylene goods are mostly made using crude oil as a basic source. Using recycled polyethylene as a raw material decreases reliance on petroleum. The production methods are energy-intensive, mainly because transforming PE into fibres involves highly heated operations that produce harmful compounds and necessitate a significant amount of energy and water. Because most PE goods are manufactured in Asia, the transportation implications of their worldwide export in terms of CO_2 emissions and direct pollution are primarily tied to sea and air freight transportation (Fig. 9.4). Since polyethylene is not biodegradable and can last for millennia in landfills, it is advised that old sheets be reused and recycled. Extra caution should be exercised when dumping in well-managed waste sites (Albadra et al. 2018).

9.2.8 *Thatching Material*

Palm leaves, reeds, and grasses are obtained from natural or cultivated plants as thatching material. Harvesting has the potential to destroy natural ecosystems and create erosion. Thatching material is gathered and processed on a domestic or small-scale industrial scale. If not correctly handled, its processing might result in water contamination. Processing thatch does not involve the use of extra materials or energy sources. Thatching methods can aid in developing community-led and traditional livelihood options while contributing to good grassland management. Only thatching materials readily available locally and do not harm the environment should be utilized. Because the material must be seasoned, thatch cannot be used all year. Thatch should be treated to ensure the goods' long-term durability (Venkatarama Reddy 2009) (Fig. 9.5).

Fig. 9.4 Emergency shelter tent built with plastic sheet

Fig. 9.5 Thatched roof house

9.2.9 Cement or Concrete

Limestones and other minerals mined from quarries or mines make cement. Mining may have severe social and environmental consequences due to its extraction. Limestone extraction can modify the pH value/alkalinity of the soil, affecting hydrogeology and the host community's livelihoods. When cement is used to mix concrete, mortar, or plaster, it is combined with other components such as quarry material and water. Cement is manufactured on a massive scale, and the wastes that arise have significant environmental consequences. Cement manufacturing is responsible for around 5% of worldwide CO_2 emissions as it needs a considerable amount of raw materials and a lot of energy, which results in carbon emissions and severe air pollution and dust if fossil fuels are employed. Clinker is the most essential cement component, and its manufacture requires a lot of resources. Every year, the cement production process generates millions of tons of waste cement kiln dust, posing respiratory and pollution health problems (Fig. 9.6).

Instead of in situ mixing and optimal concrete mix/ratios, prefabricated concrete objects and premixed concrete are recommended to reduce waste during construction. To minimize damage and degradation, cement should be stored in a dry area at appropriate stack heights and never disposed of in the natural environment. Concrete constructions require less maintenance, have a longer lifespan, and provide superior protection and safety in a fire. Cement-based parts and structures should always be examined for reuse or restoration (Swamy 2015).

Fig. 9.6 Cement built house

9.2.10 Cement Blocks

Cement blocks require cement and materials from quarries and mines (sand, rock chips, and gravel). Either driven or manually operated, pressure moulds are used to make cement blocks. Dust, noise, and silt difficulties might occur in casting regions, necessitating a large amount of drying space. If the process is not well-managed, large amounts of water are required and discharged into the earth. The use and transportation of cement may negatively influence the environment. Crushed old blocks can be used as filler in Portland cement (Zahra et al. 2021).

9.2.11 Quarried Materials

Blasting is used to extract quarry materials (i.e. sand, rock chips, and gravel), which results in noise, dust, pollution, and habitat devastation. Landslides and hydrogeological consequences can occur as a result of haphazardly managed rock quarrying. Quarried material extraction also has the potential to destroy natural habitats and agriculture, contaminate water bodies, produce ponds where disease vectors can thrive, change the local hydrological regime, and cause soil erosion. After extraction, quarry material is frequently processed further to ensure that the aggregate size is consistent. In certain facilities, the material is crushed and uniformed, resulting in noise and air pollution and silting/pollution of water bodies. Local roads are also impacted by transportation (Zhao et al. 2020).

9.2.12 Inflated Materials for Flood-Affected Temporary Shelters

Inflatable materials offer a practical and efficient solution for creating temporary shelters in flood disaster scenarios. These materials are lightweight, easy to transport, and can be rapidly deployed, providing immediate protection and comfort to refugees. One of the most commonly used materials for inflatable shelters is PVC-coated polyester fabric. This material is known for its durability, waterproof properties, and resistance to UV rays, making it ideal for harsh weather conditions. PVC-coated polyester can be used to construct the outer shell of inflatable tents, ensuring that the shelter remains dry and secure even in heavy rain (Humanitarian Library 2020).

Another excellent material for inflatable shelters is thermoplastic polyurethane (TPU). TPU is highly elastic, abrasion-resistant, and flexible, which makes it perfect for creating airtight chambers within the shelter. These chambers can be inflated to provide structural support, ensuring that the shelter remains stable and robust. TPU can also be combined with other materials to enhance the overall strength and longevity of the shelter. For instance, TPU can be used in the construction of inflatable beams and support structures, providing a solid framework that can withstand strong winds and other environmental stresses.

Nylon with a silicone coating is another material that is highly suitable for inflatable shelters. This combination offers a lightweight yet strong fabric that is resistant to water and mildew. The silicone coating adds an extra layer of waterproofing, making it particularly useful in flood-prone areas. Nylon with silicone coating can be used to create inflatable walls and barriers around the shelter, preventing water ingress and ensuring that the interior remains dry and habitable. These barriers can be quickly deployed and inflated, providing an additional layer of protection against rising floodwaters.

In terms of practical applications, inflatable tents made from PVC-coated polyester and TPU can be rapidly set up using manual or electric pumps. These tents offer immediate shelter and can be anchored to withstand strong winds and heavy rain. Inflatable rafts and platforms made from TPU and nylon with silicone coating can be used to transport people and supplies in flooded areas, providing a stable and buoyant platform that ensures safe passage. Additionally, inflatable walls and barriers can be deployed around shelters to prevent water ingress, offering enhanced protection and safety for the inhabitants (UNHCR 2024).

By utilizing these materials and designs, inflatable shelters can provide a versatile and effective solution for flood disaster response, ensuring that affected populations have access to safe and secure temporary housing. The combination of durability, flexibility, and ease of deployment makes these materials ideal for creating shelters that can withstand the challenges posed by flood disasters.

9.3 Newly Introduced "Green" Materials

Commonly, the products considered most sustainable are those that use natural local resources or recycled ingredients in place of virgin ones, recycle any waste generated by their manufacturing processes, have low embodied energy, are biodegradable, have minimal derivation from petrochemicals, and are designed for reuse and recyclability. They are also non-toxic, nonhazardous, recyclable or reusable, locally obtained, energy and water-efficient, durable, and environmentally responsible in manufacturing, occupancy, and demolition. Some green building products are presented below.

Shigeru Ban's creative application of recycled cardboard materials in the Paper Log House case study (Sect. 9.5) exemplifies the possibilities of unusual materials in the building. He demonstrated how even common materials like cardboard might be creatively and scientifically converted into useful, sustainable construction components.

9.3.1 Insulated Concrete Forms (ICFs)

Insulated Concrete Forms (ICFs) are left in place after the concrete is poured for a foundation or wall. The forms are generally made of foam insulation, such as expanded polystyrene (EPS), and are filled with reinforced concrete to create a solid structure. The sustainable benefits of ICF systems can be summarized in five points: optimized energy performance, durability, recycled material content, local materials, and improved indoor air quality. Energy efficiency is due to the thermal mass of concrete, which helps absorb and release heat slowly, and the system's air tightness and continuous insulation. Other benefits provided by the solid structure include sound insulation, resistance to fire, pests, storms, and high winds (Lewis 2012).

9.3.2 Structural Insulated Panels

Structural insulated panels have a core of rigid foam insulation between an exterior and interior skin. Panels are custom-designed, prefabricated, and assembled on-site, which reduces waste, construction time, and labour costs. Their solid structure allows air tightness with high thermal insulation and minimum levels of air infiltration. In addition, this integrated system uses less wood than a conventional wood-frame house (Mugahed Amran et al. 2020).

9.3.3 Engineered Wood

A sustainable alternative to traditional wood siding or joists is engineered wood, made of wood fibres combined with bonding agents creating a composite lightweight product with superior strength. Using engineered boards for flooring and roof sheathing instead of common plywood can lead to savings in embodied energy (Winchester and Reilly 2020).

9.4 Materials for Healthy Living

Indoor air quality (IAQ) ensures occupants' health and comfort. Lack of daylight, poor ventilation, excessive noise, or noxious compounds emitted from building materials can have lasting consequences on health and may result in allergies, stress, or the transmission of communicable diseases. Comfort is a complex issue affected by factors such as type of activity, clothing, age, gender, air temperature, movement, radiant temperature, humidity, noise, light, and odours.

Some toxic chemicals and materials are used in construction, finishes, and consumer goods. Some of these products pollute indoor air or water supplies, while others cause damage by contact. These can affect the factory workers who manufacture them, the people who install them, the occupants of the new building, and construction workers who renovate, refurbish, or demolish the building at the end of its useful life (Banerjee et al. 2021).

Two factors determine the number of pollutants emitted into the air. The first is the quantity of the material used, and the second is the number of pollutants the material will emit over its useful life. Some types of emissions are affected by changes in environmental conditions, such as temperature and humidity. Others may be affected by air currents (Winchip 2011).

Some products, such as carpets, particleboard, and other manufactured wood products, are subject to standard testing procedures, which can be compared. Manufacturers of these products who do not disclose full emission test reports should be considered with skepticism, if being considered at all. In cases where testing is not common or standardized, the manufacturer should disclose the product's contents and a description of the manufacturing processes and treatments likely to affect its emissions. The willingness of manufacturers to take steps to improve a product's performance in reducing harmful emissions should also be considered in the selection process. For instance, pretreatment and aeration of a product prior to its installation in a building could significantly improve that product's IAQ performance.

It should also be mentioned that considerations for good IAQ are not limited to residential construction. Many furnishings are made of particleboard products or have lacquered finishes with high emissions levels, sometimes for lengthy periods. While furniture selection is usually beyond the responsibility of either the architect or the builder, it might be helpful to include suggestions in an owner's manual, along with recommendations for cleaning products and maintaining ventilating equipment (Green Building 2017).

The most harmful indoor contaminants emitted from building materials can usually be traced back to a few chemical components. Most harmful emissions originate from adhesives, products that require polymerization or foam expansion, vinyl products, particularly those with plasticizers, and maintenance products such as waxes and cleaning agents.

9.4.1 Materials for Positive Social Effects

Important social impacts that contribute to the well-being of communities and the environment are brought about by the utilization of sustainable materials in the building industry, particularly in the context of homes and shelters constructed from these materials. Communities can reap the benefits of healthier living conditions, reduced environmental impact, and long-term social and economic stability if they incorporate sustainable materials into their construction projects. The enhancement of public health is one of the most important social impacts that can be attributed to the utilization of sustainable resources. It is possible to create healthier living circumstances for tenants by using materials that are non-toxic, energy-efficient, and ecologically friendly. These materials decrease exposure to dangerous chemicals, enhance air quality, and manage temperatures inside the building. The importance of this cannot be overstated when it comes to vulnerable groups, such as those living in communities with low incomes or those who have been relocated, whose health problems are frequently made worse by having bad living conditions.

Moreover, the use of sustainable resources helps to strengthen the economic and social stability of communities. Communities have the ability to provide employment opportunities and boost their economies by utilizing renewable resources or obtaining materials from local sources. The development of a sense of pride and ownership in local initiatives, as well as the encouragement of community engagement in the construction and maintenance of buildings, are all factors that contribute to the promotion of social cohesion. In addition, sustainable materials are typically more long-lasting and energy-efficient, which results in lower long-term expenditures for residents. This can help residents feel less stressed about their finances and improve their quality of life.

9.5 Case Study: Paper Log Houses by Shigeru Ban

For a long time, Shigeru Ban has been admired for his work with unconventional structural systems and creativity. He has conducted groundbreaking research using recycled cardboard materials, observing many flexible shapes and ways of construction. Ban also made it feasible for recycled cardboard materials to resist water and fire. The "Invisible Structure (where the exterior materials hide the structure)", according to Shigeru Ban, is one of the most critical design techniques. He is focused on creating architecture that benefits people and proving that paper, like steel or concrete, can be a viable construction material. He takes a straightforward and cost-effective approach to dealing with the complexity of post-disaster issues. When comprehending Ban's design philosophy, these two factors are critical (Rethinking the Future n.d.) (Fig. 9.7).

A devastating earthquake struck Kobe, Japan, in 1995. It wreaked havoc on the city, displacing over 300,000 people who were left destitute and homeless. More than 2000 structures in Kobe were destroyed due to the quake. Recognizing the crucial demands 6 months after the disaster, architect Shigeru Ban designed a unique technique to build these emergency shelters in response to the people's dire needs. The Paper Log House was the name given to this unique building prototype.

The Paper Log House comprises numerous essential components that address the needs of earthquake sufferers. It is cost-effective and may rely on the supply of readily available materials. Because there is excess material, the victims or the government may quickly get the resources needed to construct the structures. Moreover, because the structure's design and construction are simple, they may be erected by

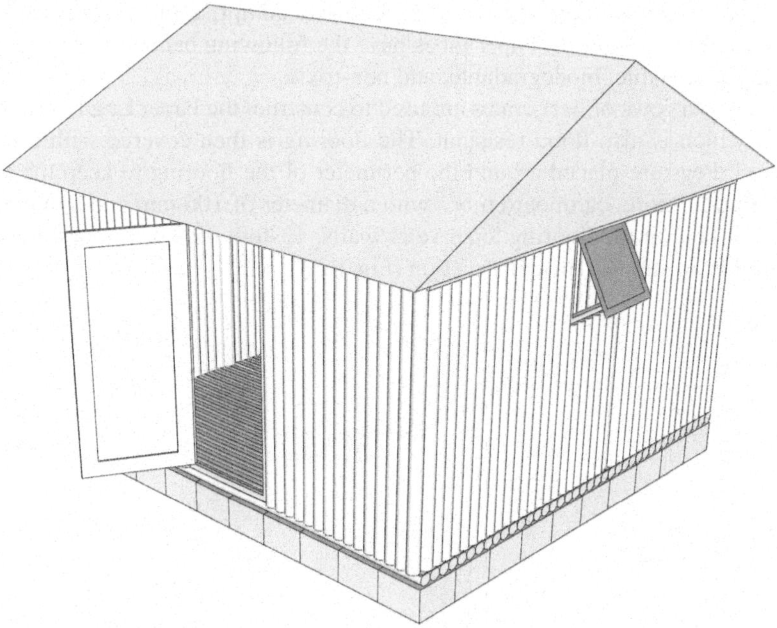

Fig. 9.7 Paper Log House by Shigeru Ban

untrained labourers or, to put it another way, by anyone. Even without any machinery, the process of building this house is quickened. Eight locals can construct each Paper Log home in just 2 days when provided with all the resources. It encourages people to seek quick shelter if they experience any more earthquakes. Furthermore, Shigeru Ban‘s invention allowed these dwellings to be readily removed, allowing the government to avoid spending considerable money cleaning up the area. The pieces that have been dismantled can also be recycled. As a result, the Paper Log House demonstrates how conscious decisions about building materials can consider the needs of displaced locals, the cost-effective concerns of the governments involved, and a negligible adverse environmental impact. Shigeru Ban's Paper Log House in Kobe, Japan, was a successful concept for emergency housing. He also introduced the concept to other nations worldwide after noticing its good application. After that, earthquake survivors in Turkey and India were given Paper Log Houses to live in. Its concept was utilized to develop transitional housing for refugees in Kosovo and to build for individuals displaced by genocide in Rwanda. Paper Log Houses quickly provided sufferers with more affordable and sanitary accommodation. Paper Log House's significance lies in its technological innovation and the humanitarian issues it addressed (Carbonari and Librelotto 2018).

Construction Process

Shigeru Ban's use of simple, recyclable materials allowed him to create a functional disaster relief shelter that was also a work of art. Beer crates, sandbags, plywood, cardboard tubes, waterproof tape, and steel rods comprise the 13-foot-by-13-foot shelter Paper Log Houses. Paper tubes have the following benefits: they are sturdy, recyclable, reusable, biodegradable, and non-toxic.

About four rows of beer crates are used to construct the Paper Log House's foundation, which is also flood resistant. The flooring is then covered with a layer of plywood. Pegs are placed around the perimeter of the flooring to keep the floor in place. Furthermore, cardboard tubes with a diameter of 106 mm and a thickness of 4 mm are laid on the flooring to serve as walls; 13-foot rods were put horizontally in the tubes to support the construction (Fig. 9.8).

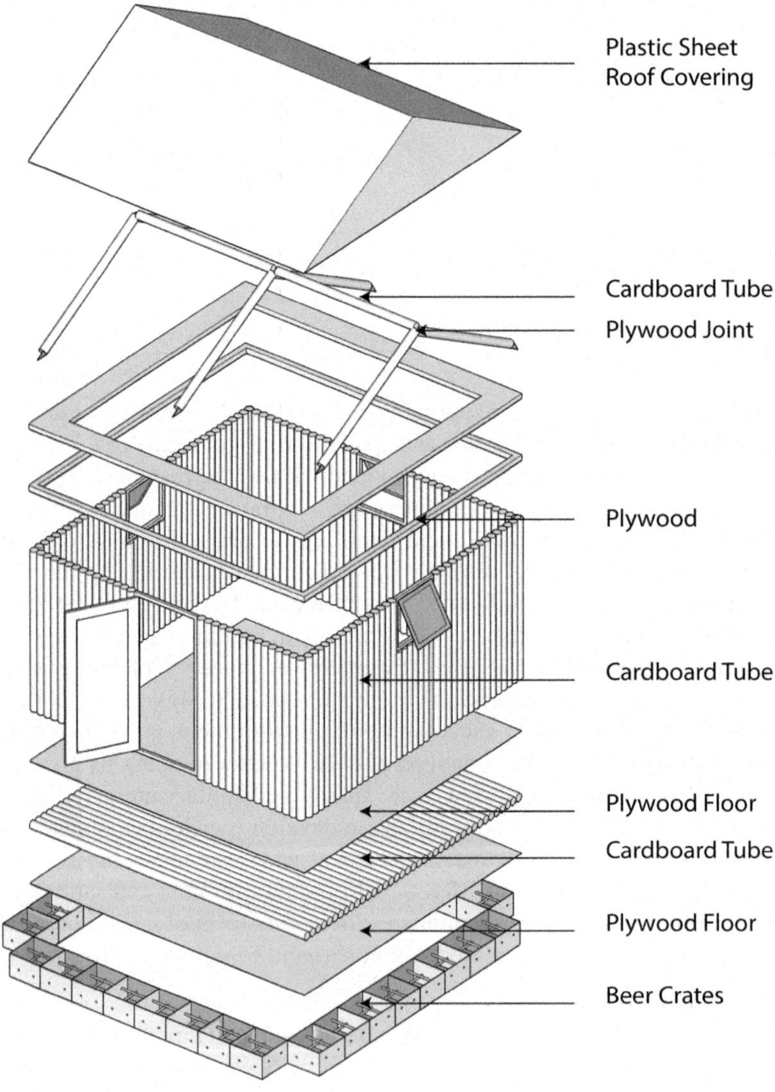

Fig. 9.8 Axonometric diagram showing the construction detailss

As part of the roofing system, a plywood frame is bonded to the frame with tent-like material. The roofing system is mechanized, allowing it to operate under various weather situations. It is handled as a distinct element from the rest of the shelter, allowing air to circulate in the summer and warm air to circulate in the winter. Furthermore, the waterproof tape and sponges keep the weather out of the temporary shelter (Rethinking the Future n.d.).

In contrast to the resident's former disaster-stricken house, Shigeru Ban observed all the possibilities to make the shelter pleasant and economical. The shelter is simple to put together and takes less than 6 h to complete. The Paper Log House is of great value, with a total cost of under US $2000. The supplies, including paper tubes that may be produced on-site, are simple to store and transport to the job site (McQuaid and Ban 2006).

9.5.1 Lessons Learned

- **Low-Carbon Footprint Materials**
- Building materials should have a low-carbon footprint to mitigate the contribution of shelters to climate change. Designers must prioritize materials with lower embodied energy, such as bamboo, adobe bricks, and timber, which are sustainable and less energy intensive compared to conventional materials like concrete and steel.
- **Ecological Considerations**
- The ecological footprint of building materials—from extraction and processing to transportation and disposal—must be minimized. Using local, renewable, or recycled materials not only reduces the carbon footprint but also promotes circular economy practices in shelter design.
- **Durability and Life Cycle Impact**
- The entire life cycle of materials—harvesting, manufacturing, installation, and eventual dismantling—should be considered to ensure that shelters remain durable and eco-friendly over time. Materials like clay, adobe, and prefabricated concrete forms, for example, have varying durability and environmental impacts, and selecting materials with a long life cycle can lead to more sustainable refugee settlements.
- **Adapting to Climate Conditions**
- Shelters must be designed with climate resilience in mind, ensuring they can withstand local environmental conditions. The choice of materials plays a critical role in this; for example, selecting materials with high thermal mass in hot climates or weather-resistant materials like corrugated galvanized iron (CGI) sheets in regions prone to heavy rain or wind can make a significant difference in long-term shelter performance.
- **Recycling and Reusability**
- The potential for reusing or recycling materials at the end of a shelter's life cycle should be considered from the design stage. For instance, materials such as CGI

sheets, bamboo, or timber can be repurposed for future shelter construction or other community needs, reducing waste and promoting sustainability.

Weblink of Case Study Project: https://shigerubanarchitects.com/works/paper-tubes/paper-log-house-india/

9.6 Conclusion

Addressing the challenges of shelter construction in refugee camps within the context of climate change necessitates a paradigm shift towards sustainability. The use of environmentally responsible materials, low-carbon construction methods, and sustainable design principles are crucial in minimizing the ecological footprint of shelters while ensuring their durability and functionality. The key to sustainable refugee shelters lies in the careful selection of materials with low embodied energy, the promotion of local and renewable resources, and the integration of innovative green technologies. These practices reduce greenhouse gas emissions and contribute to the well-being of displaced populations by improving indoor air quality and creating safer, healthier living environments.

Despite positive advancements in developing low-emitting building materials, shelter organizations remain far from achieving an adequate standard for refugee shelters. This may be due to increased prices for sustainable building materials or a lack of awareness of newer products. Many refuge shelters often overlook the use of environmentally conscious building products due to cost, failing to realize that maintenance and future costs will be reduced. In the future, sustainable products must be part of the shelter construction process. Hence, a sustainability assessment for building materials is needed while choosing shelter materials. Combining initiatives for greener shelters with effective planning policies will promote a sustainable living environment in the refugee camp.

Discussion Questions

- What are the major criteria for selecting materials for refugee housing, and why are they important?
- How can the use of locally available materials and equipment benefit the construction of emergency shelters?
- Why is the ability to reuse materials an important consideration for sustainability in emergency shelters?
- What are the advantages of using lightweight materials that are easily transportable in a shipping container for shelter construction?
- What specific material considerations are necessary for shelters in climates with extended periods of cold weather, rain, and snow?

References

Albadra D, Coley D, Hart J (2018) Toward healthy housing for the displaced. J Archit 23(1):115–136. https://doi.org/10.1080/13602365.2018.1424227

Banerjee A, Melkania NP, Nain A (2021) Indoor air quality (IAQ) in green buildings, a pre-requisite to human health and well-being. Digital Cities Roadmap, pp 293–317. https://doi.org/10.1002/9781119792079.CH9

Carbonari LT, Librelotto LI (2018) Temporary housing made from recycled paper tubes. INCREaSE, pp 503–512. https://doi.org/10.1007/978-3-319-70272-8_39

Chettri N, Gautam D, Chikermane S, Prakash V, Vaghela K (2021) Sustainability assessment of Bhutanese vernacular wattle and daub houses. Innov Infrastruct Solut 6(4):1–13. https://doi.org/10.1007/S41062-021-00576-Z

Dabaieh M, Emami N, Heinonen JT, Marteinsson B (2020) A life cycle assessment of a 'minus carbon' refugee house: global warming potential and sensitivity analysis. Archnet-IJAR 14(3):559–579. https://doi.org/10.1108/ARCH-11-2019-0258/FULL/XML

Green Building (2017) Indoor air quality. http://www.greenbuilding.com/indoor-air-quality/. Accessed 12 Jun 2022

Humanitarian Library (2020) Flood resilient shelter reconstruction, retrofitting and repair technical guidelines

Hussain J, Khan A, Zhou K (2020) The impact of natural resource depletion on energy use and CO2 emission in Belt & Road Initiative countries: a cross-country analysis. Energy 199:117409. https://doi.org/10.1016/J.ENERGY.2020.117409

Joglekar SN, Kharkar RA, Mandavgane SA, Kulkarni BD (2018) Sustainability assessment of brick work for low-cost housing: a comparison between waste based bricks and burnt clay bricks. Sustain Cities Soc 37:396–406. https://doi.org/10.1016/J.SCS.2017.11.025

Lewis D (2012) Use of insulating concrete forms in residential housing construction 85

Mancini MS, Galli A, Niccolucci V, Lin D, Bastianoni S, Wackernagel M, Marchettini N (2016) Ecological footprint: refining the carbon footprint calculation. Ecol Indic 61:390–403. https://doi.org/10.1016/J.ECOLIND.2015.09.040

McEwan A, Marchi E, Spinelli R, Brink M (2020) Past, present and future of industrial plantation forestry and implication on future timber harvesting technology. J For Res (Harbin) 31(2):339–351. https://doi.org/10.1007/S11676-019-01019-3/TABLES/1

McQuaid M, Ban S (2006) Shigeru Ban. Phaidon Press, London, New York, NY

Mugahed Amran YH, El-Zeadani M, Huei Lee Y, Yong Lee Y, Murali G, Feduik R (2020) Design innovation, efficiency and applications of structural insulated panels: a review. Structure 27:1358–1379. https://doi.org/10.1016/J.ISTRUC.2020.07.044

Rethinking The Future Paper Log Houses by Shigeru Ban: Recyclable Materials and Architecture (n.d.). https://www.re-thinkingthefuture.com/case-studies/a4922-paper-log-houses-by-shigeru-ban-recyclable-materials-and-architecture/. Accessed 12 Jun 2022

Riaz MH, Khitab A, Ahmed S (2019) Evaluation of sustainable clay bricks incorporating brick kiln dust. J Build Eng 24:100725. https://doi.org/10.1016/J.JOBE.2019.02.017

Salih MM, Osofero AI, Imbabi MS (2020) Critical review of recent development in fiber reinforced adobe bricks for sustainable construction. Front Struct Civil Eng 14(4):839–854. https://doi.org/10.1007/S11709-020-0630-7

Swamy RN (2015) Holistic design: key to sustainability in concrete construction. Proc Inst Civil Eng Struct Build 146(4):371–379. https://doi.org/10.1680/STBU.2001.146.4.371

Tu C, Suweis S, D'Odorico P (2019) Impact of globalization on the resilience and sustainability of natural resources. Nat Sustain 2(4):283–289. https://doi.org/10.1038/s41893-019-0260-z

UNHCR (2021) Shelter and sustainability

UNHCR (2024) Emergency shelter solutions and standards. https://emergency.unhcr.org/emergency-assistance/shelter-camp-and-settlement/shelter-and-housing/emergency-shelter-solutions-and-standards. Accessed 2 Oct 2024

van der Lugt P, van den Dobbelsteen AAJF, Janssen JJA (2006) An environmental, economic and practical assessment of bamboo as a building material for supporting structures. Constr Build Mater 20(9):648–656. https://doi.org/10.1016/J.CONBUILDMAT.2005.02.023
Venkatarama Reddy BV (2009) Sustainable materials for low carbon buildings. Int J Low-Carbon Technol 4(3):175–181. https://doi.org/10.1093/IJLCT/CTP025
Winchester N, Reilly JM (2020) The economic and emissions benefits of engineered wood products in a low-carbon future. Energy Econ 85:104596. https://doi.org/10.1016/J.ENECO.2019.104596
Winchip SM (2011) Sustainable design for interior environments. Fairchild Books, New York, NY
Woodard AC, Milner HR (2016) Sustainability of timber and wood in construction. Sustainability of Construction Materials, pp 129–157. https://doi.org/10.1016/B978-0-08-100370-1.00007-X
Zahra T, Thamboo J, Asad M (2021) Compressive strength and deformation characteristics of concrete block masonry made with different mortars, blocks and mortar beddings types. J Build Eng 38:102213. https://doi.org/10.1016/J.JOBE.2021.102213
Zea Escamilla E, Habert G (2014) Environmental impacts of bamboo-based construction materials representing global production diversity. J Clean Prod 69:117–127. https://doi.org/10.1016/J.JCLEPRO.2014.01.067
Zhao Y, Qiu J, Xing J, Sun X (2020) Recycling of quarry dust for supplementary cementitious materials in low carbon cement. Constr Build Mater 237:117608. https://doi.org/10.1016/J.CONBUILDMAT.2019.117608

Chapter 10
Charting a Path Forward

Keyword Sustainable emergency shelters · Refugee housing · Shelter design · Displaced communities · Planning and design steps · Viable solutions for refugee shelters · Economic potential of migration · Social challenges of migration · Climate change and refugee crisis · Crisis response solutions

This chapter outlines the fundamental steps necessary for the successful planning and implementation of sustainable refugee shelters. Each of the preceding chapters focuses on a subject vital to the structure of sustainable emergency shelters and communities. Ideally, planners and designers should include all of them in their designs. However, refugee housing communities in many countries do not provide all the facilities and shelters with sustainable ideals; thus, compromises are often made. Although displaced people tend to live in these communities for years, they live without all the necessities of human life. This concluding chapter reviews the practical aspects of building sustainable emergency shelters. It describes basic steps that planners can follow at the outset of their design, offers an outlook on refugee shelter design and construction, and outlines solutions for these growing crises.

10.1 Basic Initial Steps

One challenge planners of new sustainable refugee shelters face at the outset of a design process are integrating the various elements outlined in this book's chapters. Following is the summary of critical considerations described in the previous chapters. When conflict happens in a country due to civil war or other political reasons, it needs an immediate response to provide shelter for the displaced people. It takes a lot of time to plan and make decisions regarding the design of shelters and site planning. In some cases, the planning and decision-making processes take more time than the construction of shelters. Therefore, planners, architects, designers, and engineers should follow an integrated design process to house the refugees immediately.

A. Friedman, B. Chaki, *Fundamentals of Planning and Designing Sustainable Post-Disaster Shelters*, https://doi.org/10.1007/978-3-031-83317-5_10

Furthermore, the designers must establish clear goals and consider a set of criteria to develop a creative and effective solution for the long-term stability and sustainability of the shelters. Apart from these, during the planning process, the designers need to include the displaced community to understand their family sizes, way of life, needs, and challenges in the camp. Humanitarian organizations responsible for ensuring the well-being of the refugees can conduct stakeholder meetings among the design team, community leaders, and stakeholders to discuss their needs, issues, and concerns regarding their shelters. Additionally, the organizations may consider developing metrics to measure success after constructing the prototype shelters for the refugees. Generally, the construction of pilot shelters helps explore practical solutions. By comparing the metrics and goals achieved in the pilot shelters, new goals can be set, and the final design decision can be taken accordingly.

The following flow chart shows the steps to be followed to design housing for refugees (Fig. 10.1):

10.1.1 Step 1: Designing the Master Plan of the Camp

After any conflict, the shelter organization's first responsibility is to make the overall site plan of the camp. First, they need to count the number of beneficiaries, their household size, and the number of children and elderly in the family. Then, they can start designing the master plan, considering climate and topography, accessibility, natural elements, amenities and infrastructure, and modularity and adaptability.

10.1.2 Step 2: Determining the Area and the Number of Rooms in the Shelter

It is recommended that the size of the shelter be determined based on the household size of the refugee families. The shelter organization should have a survey on the number of people, the number of children in the family, and their age group. The interior area of the shelters should be determined considering the minimum area per person and household sizes of the refugee families.

10.1.3 Step 3: Defining the Way of Sanitation and Water Collection

The camp design should have access to water supply and toilets. Many refugee camps, including the Rohingya refugee camp in Bangladesh, have common water supply points. However, while designing, it should be kept in mind that the water

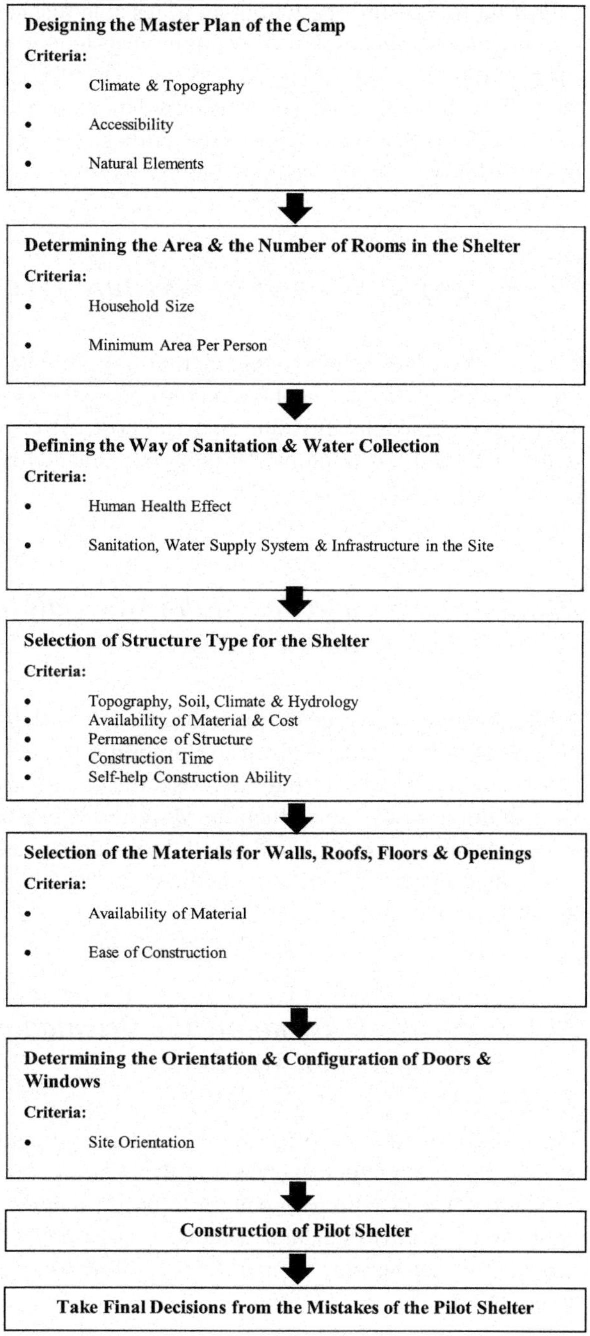

Fig. 10.1 Steps showing the planning and decision-making process

collection points must be within an accessible distance and sufficient, considering the number of refugees. For the sanitation system of the refugees, pod toilets are an easy solution for self-help construction due to their easy and quick installation process. These toilets can also be transferred quickly, and they solve many of the common problems that the refugees face in the refugee camp, such as hygiene issues, inaccessibility, and inefficient and faulty fixtures.

10.1.4 Step 4: Selection of Structure Type for the Shelter

For any type of refugee housing, including Rohingya refugees, the primary criteria for selecting materials for the structure and exterior envelope are durability, construction time, availability, and material cost. Also, for self-help construction, the material should have compatibility to be managed by the refugees themselves to avoid unwanted injuries.

10.1.5 Step 5: Selection of the Materials for Walls, Roofs, Floors, and Openings

While selecting materials for the interior, such as walls, floors, doors, and windows, prior attention should be given to durability, protection from an adverse climate, construction time, and availability. Also, the materials should be prefabricated in the workshop for ease of construction. Moreover, if the shelter kits are transported by the organization, prefabricated wood floors, timber-framed doors, and windows are sustainable materials for the immediate construction, ensuring the permanence of the structure.

10.1.6 Step 6: Determining the Orientation and Configuration of Doors and Windows

The orientation of the shelter and the openings should be designed considering the passive design strategies, such as how the openings should be placed considering the wind direction and sun path pattern to utilize natural ventilation, heating, and daylight. In every environmental context, it is essential to design a comfortable residence considering air temperature, airflow direction, relative humidity, and precipitation.

10.1.7 *Step 7: Construction of Pilot Shelter and Taking Final Decisions from the Mistakes of the Pilot Shelter*

After the design, a pilot shelter needs to be built. During construction, the designers and contractors need to observe whether the construction process can be a self-help construction process implemented by refugees and evaluate the construction time. From the pilot shelter construction, the planners and architects can evaluate and refine the design for final construction. The pilot shelter construction allows the designers to make the best design for emergency needs.

10.2 Ensuring Viable Solutions

Implementing sustainability in refugee housing presents several challenges and often involves compromises. Some strategies that aim to improve the design's environmental quality may be either too costly or require changes that the project's initiator might resist implementing. Therefore, a requirement to keep in mind is the existing building industry's efficiency, which has evolved into a concise, unique operation. Any innovation that severely delays this process runs a high risk of being rejected or will prolong the process substantially. Designers should thoroughly assess the impact of their proposal on standard practices before attempting to modify them.

From the resident's perspective, the shelter is their new home where they might stay for many years. Therefore, a designer must consider cultural, environmental, and other aspects. The end user might reject innovative concepts if their needs are not considered. Economic considerations often take precedence in any development process. It is not sure how much more the shelter organization would be willing to pay for an environmentally sound alternative. The shelter organizations would most likely prefer to invest in functional amenities, such as bigger shelters for more people, rather than net-zero technologies. Even if the added investment would decrease energy costs, the expected occupancy period may not justify ideas with long payback periods.

Developing a sustainable emergency housing community involves a process that includes site planning, design, and construction. As noted in Chap. 2, when examining the environmental impact and comparing alternatives, the project's entire life cycle should be accounted for. The efficiency with which natural resources, including energy, are utilized in the shelters and the community at macro and micro levels will determine the project's overall environmental impact. For obvious reasons, not all environmental challenges can be solved in an absolute sense. What may appear to be a suitable alternative in one respect, such as having solar power, may exhibit compromising qualities in another, such as using non-renewable natural resources. Without reliable, quantitative data, many decisions are often based on qualitative assessments, some of which are objective. Although life cycle analyses are complex, they can be helpful in choosing alternatives.

The case studied in the earlier chapters proves the necessity of sustainable and environment-friendly design strategies for refugee housing and shelter design. The Tukuraki Village case in Fiji demonstrates that after the cyclone in 2012, by adopting the site selection principles effectively, the Fijian government obtained safe housing for the community. Chapter 5, a very recent case study in Ukraine, shows how a temporary housing system can give internally displaced Ukrainians a respectable place to live. The project considered many functional aspects of the design (i.e. modularity, scaling possibility, and flexibility) to make the shelter design sustainable. In the Rohingya Refugee Camp case in Bangladesh, the project was designed to embrace a "whole settlement approach," incorporating site management, infrastructure, and amenities. This helped to minimize any negative environmental impact, which was the foremost priority of the project at all stages. The Paper Log Houses case study described creative and sustainable shelter materials solution developed by renowned architect Shigeru Ban. He has undertaken ground-breaking research using recycled cardboard, noting several adaptable building techniques. The Prefab Housing Case in Japan in Chap. 8 shows that prefabricated shelters could be a rapid and cost-effective solution to refugee housing. The Hexagonal Shelter Cluster case in Chap. 9 is an excellent example of a generative housing cluster design for the refugee camp. These case studies elucidate the importance of applying sustainable design strategies from the predesign stages, especially because when it comes to refugee camp design, the governments often ignore their needs as human beings.

10.3 Economic Potential and Social Challenges of Migration

Despite many obstacles, migrant and refugee populations can significantly contribute to the Gross Domestic Product (GDP) of receiving nations. Following are some ways they can contribute to the economic growth of the host country:

- **Participation in the Labour Force**: Migrants frequently fill labour market voids that native populations might not be able or willing to fill. Their varied backgrounds and skill sets can boost output in several industries. The GDP grows because of the higher output resulting from this increased labour force participation.
- **Entrepreneurship**: A lot of immigrants launch enterprises in the nations where they migrate. These companies boost the economy, produce revenue, and create jobs. The introduction of novel concepts, goods, and services by migrant entrepreneurs frequently stimulates innovation and competitiveness in the economy of their host nation.
- **Demand and consumption**: Population migration fuels the demand for products and services in the economy. Businesses benefit from these higher sales and revenue due to the increased consumption, which further boosts economic growth.

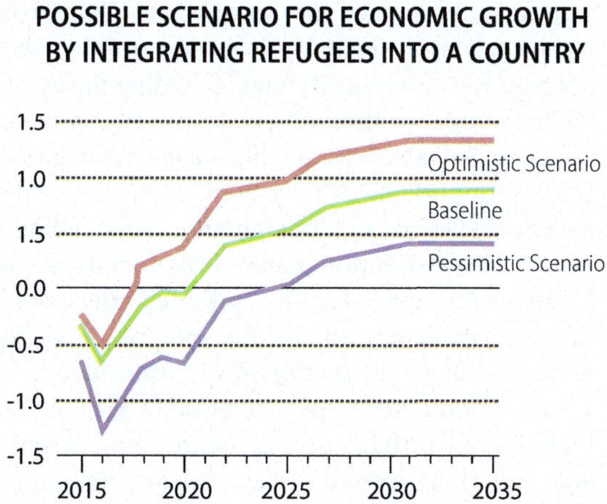

Fig. 10.2 Possible scenario for economic growth by integrating refugees into a country

- **Cultural diversity and innovation**: As refugees and migrant populations contribute to the cultural fabric of host nations, cross-cultural dialogue and inventiveness increase. Because diversity promotes fresh viewpoints and methods of approaching problems, it can stimulate creativity and propel economic progress. Cultural exchange can also draw tourists and foreign business prospects, increasing GDP even more.

According to estimates from the German Institute of Economic Research, Germany's acceptance of more migrants than any other European nation would directly result in a.5% GDP gain after ten years (Fratzscher and Junker 2015). Even in the most pessimistic scenario, according to their analysis (Fig. 10.2), Germany's economy will benefit net from the influx of migrants throughout that ten-year period. The rise will originate from three primary sources: the diversity of immigrants, the beneficial impact of refugees on the labour market, and the rising demand brought about by refugee expenditures.

However, several things can work against the economic benefits of migration, such as prejudice, social exclusion, and ineffective government programs. As previously noted, a contradiction exists in certain receiving nations that may marginalize migrants, preventing them from assimilating and engaging in the social and economic life of the community. For instance, the restriction of refugees in camps on islands, as seen in Bangladesh (details in Sect. 3.6) and other locations, makes it more difficult for them to make a significant economic contribution. Such actions isolate immigrants and prolong cycles of poverty and dependency rather than integrating them into society and the labour market. In summary, migrant and refugee populations can make substantial contributions to GDP and economic growth in host nations; however, to fully realize these advantages, systemic integration barriers must be addressed, and inclusive policies that value the diverse skills and contributions of migrants must be promoted.

While this book delves into the technical intricacies of designing and constructing refugee housing, it is crucial to also consider the social dimensions that significantly impact the success of such projects. Socioeconomic factors, for instance, play a pivotal role in determining the affordability and accessibility of housing for refugees. By understanding the economic constraints and opportunities within refugee communities, designers can create solutions that are not only cost-effective but also sustainable in the long term. Socio-environmental aspects are equally important, as they address the integration of housing projects within the local ecosystem and community. This includes considering the environmental impact of construction materials and methods, as well as ensuring that the housing solutions promote a healthy and safe living environment. Additionally, socio-techno factors, which encompass the integration of technology in housing design, can greatly enhance the quality of life for refugees. This might involve the use of innovative building techniques, renewable energy sources, and smart home technologies that can make living spaces more efficient and comfortable. By taking a holistic approach that includes these socio-economic, socio-environmental, and socio-techno considerations, the process of designing and constructing refugee housing can be significantly improved. This comprehensive approach not only addresses the immediate needs of refugees but also contributes to their long-term well-being and integration into society. Ultimately, the success of refugee housing projects hinges on the ability to balance technical proficiency with a deep understanding of the social context in which these solutions are implemented.

10.4 The Interconnection of Climate Change Mitigation and the Refugee Crisis

Global warming is a significant phenomenon that impacts ecosystems, economies, and civilizations everywhere. One of the most urgent problems made worse by climate change is population relocation, which gives rise to the refugee crisis. There is a crucial connection between mitigating climate change and addressing the refugee crisis. Incidents brought on by climate change, such as harsh weather, rising sea levels, and desertification, have a disastrous impact on societies and make large areas uninhabitable. People are consequently compelled to become climate refugees and leave their homes in quest of safety and a means of subsistence. As global warming continues, climate-related disasters are predicted to occur more frequently and with greater intensity, worsening the refugee situation (Fig. 10.3).

Existing social, political, and economic vulnerabilities are exacerbated by climate change, increasing the risk of conflict and instability, which in turn drive displacement. Competition for scarce resources, such as arable land and water, might, for instance, intensify tensions and lead to conflicts that force entire communities to relocate. Addressing the underlying causes of displacement and averting other humanitarian catastrophes depend heavily on mitigating climate change. The intensity of climate-related disasters can be lessened, and vulnerable communities

Fig. 10.3 Floods and droughts are common phenomena due to climate change

can be shielded from displacement by cutting greenhouse gas emissions and putting adaptation measures in place. Furthermore, funding programs for resilience-building and sustainable development might lessen the likelihood of community displacement by assisting them in adjusting to shifting environmental conditions. We can alleviate the adverse effects on people and communities and, in the end, solve the refugee crisis by tackling the factors that lead to climate-induced migration.

The following are actionable steps to reduce climate change:

- **Shifting towards renewable energy**: Reducing greenhouse gas emissions and addressing climate change require phasing out fossil fuels and switching to renewable energy sources like solar, wind, and hydropower. To hasten this transition, governments, corporations, and individuals must invest in the infrastructure for renewable energy sources and encourage energy-saving practices.
- **Sustainable land use and conservation**: You may help store carbon dioxide from the atmosphere and preserve biodiversity by preserving forests, recovering damaged lands, and encouraging sustainable agricultural practices. Sustainable land management techniques should be combined with conservation initiatives to improve ecosystem resilience and lessen the effects of climate change.
- **International cooperation and climate policy**: To achieve significant carbon reductions and curb global warming, national and international climate policies must be robust and well-coordinated. Governments must adopt measures to encourage climate-friendly activities across all sectors and adhere to the aggressive climate targets outlined in accordance with accords like the Paris Agreement.

- **Adaptation and resilience-building**: It is imperative to fund adaptation strategies, including early warning systems, community-based resilience projects, and infrastructure improvements to help vulnerable areas deal with the effects of climate change. Prioritizing the needs of marginalized groups and ensuring fair access to resources and assistance are essential aspects of adaptation strategies.
- **Public education and awareness**: Getting the public to support and encourage behavioural change requires spreading knowledge about the gravity of climate change and the value of group action. The goal of education and outreach initiatives should be to encourage communities and individuals to adopt sustainable lifestyles and to speak out in favour of climate action.

Due to their interconnectedness, comprehensive action is urgently needed to address both the refugee crisis and climate change mitigation. Lowering greenhouse gas emissions, supporting adaptation and resilience-building initiatives, and encouraging international collaboration can lessen the impacts of climate change and ease the suffering of displaced communities. Governments, civil society organizations, and individuals must collaborate to prioritize climate action and create a more resilient and sustainable future for everybody.

10.4.1 Lessons Learned

- **Integrated planning and community engagement**: Sustainable refugee shelters require integrated design processes that involve all stakeholders, including displaced communities. Understanding the refugees' needs, family sizes, and cultural practices is essential for creating viable and appropriate housing solutions. Early community involvement ensures that shelters are designed to reflect the realities of those living in them, enhancing long-term satisfaction and stability.
- **Modularity and flexibility**: Given the unpredictability of refugee situations, shelter designs should incorporate modularity and adaptability. This enables shelters to evolve over time, catering to changing needs and larger family sizes, while still being cost-effective and scalable.
- **Water and sanitation considerations**: Water access and sanitation are critical in refugee camps. Systems must be both practical and efficient, with solutions like pod toilets providing easy self-help construction options while maintaining hygiene standards. Proper site planning should ensure accessibility and adequacy for the camp's population.
- **Cultural and environmental adaptation**: Shelters must be designed with a deep understanding of the cultural and environmental context in which they will be placed. Residents may reject innovative ideas if they don't align with cultural norms, and shelters must be resilient to local climate conditions (e.g. ventilation, heating, and flood protection).

- **Economic potential of migrants and refugees**: Migrant and refugee populations can significantly contribute to the economy of host countries if they are integrated into the labour market. By recognizing and supporting their entrepreneurial and workforce potential, host nations can see significant economic growth.

10.5 Conclusion

In "Chap. 1: Setting the Stage: The Global Need for Shelters," we began a quest to comprehend the severe housing difficulties encountered by refugees and displaced individuals globally following conflicts and natural calamities. We observed the deficiencies of existing sheltering methods, wherein hurriedly built shelters frequently do not fulfil fundamental human requirements, leading to unlivable and unsanitary living circumstances. As we progressed into later chapters, we examined crucial elements of emergency sheltering, encompassing site selection, infrastructure development, quick delivery planning, and shelter design and materials. Every chapter provided valuable perspectives and strategies focused on enhancing the efficiency and durability of emergency shelter solutions. The concluding this chapter titled "Charting a Forward Path" reflects on the accumulated collective insight and the obstacles that await us in the future because of climate change. Effective refugee housing demands a comprehensive approach, from the initial design to the pilot shelter phase, while incorporating the unique needs of displaced populations. Engaging the community in the design process, conducting pilot projects, and continually refining based on feedback are critical for ensuring that shelters are not only functional but also culturally appropriate, safe, and environmentally responsible. Although every chapter provided crucial components for constructing enduring emergency shelters and communities, we acknowledge the disparity between recommended practices and realities in numerous refugee housing scenarios.

The adventure does not end with the completion of this book. This statement urges planners, designers, legislators, and humanitarian groups to incorporate sustainable principles into all aspects of emergency shelters. By placing the well-being and dignity of displaced populations as the top priority, we can establish a clear direction towards more robust and comprehensive approaches. As we deal with the complexities of refugee crises and natural calamities, let us stay dedicated to developing conditions where every person can have secure, respectable, and enduring housing. Collectively, we can construct a more optimistic and promising future for individuals who want assistance, guaranteeing that no one is neglected during periods of emergency.

Discussion Questions

- What are the key practical aspects of building sustainable emergency shelters that planners should consider?
- Why do many refugee housing communities fail to provide all necessary facilities and sustainable ideals?

- What compromises are often made in the design and construction of refugee shelters, and what are their impacts on displaced people?
- How can planners ensure that emergency shelters meet the basic necessities of human life for displaced people?
- What basic steps should planners follow at the outset of their design for sustainable emergency shelters?

Reference

Fratzscher M, Junker S (2015) Integrating refugees: a long-term, worthwhile investment. DIW Econ Bull 3:612–616

Weblink of Case Studies

Added Weblink of Case Study Project after the project Lessons Learned section

Chapter 2: Tukuraki Village, Fiji

Weblink of Case Study Project: https://disasterdisplacement.org/wp-content/uploads/2022/12/Information-Paper_Tukuraki_compressed-1.pdf

Chapter 3: Rohingya Refugee Transitional Housing, Bhasan Char, Bangladesh

Weblink of Case Study Project: https://fingfx.thomsonreuters.com/gfx/rngs/MYANMAR-ROHINGYA/010060Z21XP/index.html

Chapter 4: Low-cost generative hexagonal shelters cluster for refugees

Weblink of Case Study Project: https://architectsforsociety.org/our-work/hex-house-rapidly-deployable-dignified-home/

Chapter 5: Rohingya Refugee Camp, Bangladesh

Weblink of Case Study Project: https://shelterprojects.org/shelterprojects8/ref/A12-bangladesh180821.pdf

Chapter 6: Ukraine, Balbek Bureau, Kyiv, Ukraine

Weblink of Case Study Project: https://www.balbek.com/reukraine-eng

Chapter 7: Prefab Housing for Tohoku Earthquake-Affected Communities in Japan

Weblink of Case Study Project: https://www.researchgate.net/publication/276519622_A_STUDY_ON_THE_REUSE_CHARACTERISTIC_OF_WOODEN_TEMPORARY_HOUSING_-_A_study_on_wooden_temporary housing_in_fukushima_prefecture_after_the_tohoku_earthquake_-

Chapter 8: Wooden Prefabricated Temporary Housing, Aalto University Wood Program

Weblink of Case Study Project: https://www.archdaily.com/174909/liina-transitional-shelter-aalto-university-wood-program

Chapter 9: Paper Log Houses

Weblink of Case Study Project: https://shigerubanarchitects.com/works/paper-tubes/paper-log-house-india/

A. Friedman, B. Chaki, *Fundamentals of Planning and Designing Sustainable Post-Disaster Shelters*, https://doi.org/10.1007/978-3-031-83317-5

Bibliography

Aburamadan R, Trillo C, Makore BCN (2020) Designing refugees' camps: temporary emergency solutions, or contemporary paradigms of incomplete urban citizenship? Insights from Al Za'atari. City Territ Archit 7(1):12. https://doi.org/10.1186/s40410-020-00120-z

Ahmed MM, Sulaiman M (2003) Design and proper sizing of solar energy schemes for electricity production in Malaysia. In: National Power Engineering Conference, PECon 2003—Proceedings. Institute of Electrical and Electronics Engineers Inc., pp 268–271

Ajibade OO, Tota-Maharaj K, Clarke B (2017) A simplified guide to surface water drainage systems for refugee camps and internally displaced persons (IDP) temporary settlements. World Environmental and Water Resources Congress 2017: International Perspectives, History and Heritage, Emerging Technologies, and Student Papers - Selected Papers from the World Environmental and Water Resources Congress 2017: pp 519–534. https://doi.org/10.1061/9780784480595.048

Ajibade OO, Tota-Maharaj K, Hills CD, Macleod C (2019) Modelling of a sustainable refugee camp drainage system for stormwater management. Environ Sci (Camb) 5(12):2150–2161. https://doi.org/10.1039/C9EW00350A

Albadra D, Coley D, Hart J (2018) Toward healthy housing for the displaced. J Archit 23(1):115–136. https://doi.org/10.1080/13602365.2018.1424227

Al-Khatib IA, Tabakhna H (2006) Housing conditions and health in Jalazone Refugee Camp in Palestine. East Mediterr Health J 12(1-2):144–152

Alshoubaki H, Zazzara L (2020) The fragility in the land of refugees: Jordan and irrepressible phenomenon of refugee camps. J Int Stud 13(1):123–142. https://doi.org/10.14254/2071-8330.2020/13-1/8

Anderson T (2019) Transitional refugee housing. Unpublished M. Arch Thesis, University of Cincinnati, Cincinnati, OH

Angelopoulou SL (2022) Balbek Bureau develops RE: UKRAINE, a modular town system for refugees. In: Designboom. https://www.designboom.com/architecture/balbek-bureau-re-ukraine-modular-town-system-refugees-03-30-2022/. Accessed 14 May 2022

ArchDaily (n.d.) Liina transitional shelter / Aalto University Wood Program. https://www.archdaily.com/174909/liina-transitional-shelter-aalto-university-wood-program. Accessed 25 May 2022

ASHRAE (2017) ASHRAE technical FAQ. https://www.ashrae.org/File Library/Technical Resources/Technical FAQs/TC-02.01-FAQ-92.pdf. Accessed 22 May 2022

Atkinson W (2018) The orientation of buildings: or, planning for sunlight. Fotgotten Books, London

Atlas RM (1984) Sanitation and disease - health aspects of excreta and waste-water management. J Am Water Resour Assoc 20(5):803–803. https://doi.org/10.1111/j.1752-1688.1984.tb04765.x

A. Friedman, B. Chaki, *Fundamentals of Planning and Designing Sustainable Post-Disaster Shelters*, https://doi.org/10.1007/978-3-031-83317-5

Balbastre-Soldevila R, García-Bartual R, Andrés-Doménech I (2019) A comparison of design storms for urban drainage system applications. Water 11:757. https://doi.org/10.3390/W11040757

Banerjee S (2020) From Cox's bazar to Bhasan Char: an assessment of Bangladesh's relocation plan for Rohingya refugees. Observer Research Foundation

Banerjee A, Melkania NP, Nain A (2021) Indoor air quality (IAQ) in green buildings, a pre-requisite to human health and well-being. Digital Cities Roadmap, pp 293–317. https://doi.org/10.1002/9781119792079.CH9

BetterShelter (n.d.) Relief housing unit (RHU). https://bettershelter.org/rhu/. Accessed 10 Jul 2022

Bonner PC, Schmidt W-P, Belmain SR, Oshin B, Baglole D, Borchert M (2007) Poor housing quality increases risk of rodent infestation and Lassa fever in refugee camps of Sierra Leone. Am Soc Trop Med Hyg 77(1):169–175

Brandle JR, Finch S (1991) How windbreaks work. Papers in Natural Resources 121, University of Nebraska—Lincoln

Bredenoord J (2021) Self-managed cooperative housing by mutual-assistance as introduced in Central America between 2004 and 2016 the attractiveness of the 'FUCVAM' model of uruguay. J Archit Eng Technol 6:188. https://doi.org/10.4172/2168-9717.1000188

Bris P, Bendito F (2019) Impact of Japanese post-disaster temporary housing areas' (THAs) design on mental and social health. Int J Environ Res Public Health 16(23):4757. https://doi.org/10.3390/ijerph16234757

Building Safety & Resilience in the Pacific (BSRP) (2017) Tukuraki village relocation project. In: Building safety & resilience in the Pacific. https://bsrp.gsd.spc.int/wp-content/uploads/2016/09/Tukuraki-Project-Brief.pdf. Accessed 13 Aug 2021

Burden D (2000) Street design guidelines for healthy neighborhoods. Transportation research circular. Walkable Communities, Inc., High Springs, FL

Byrd N (2022) Issues in refugee adaptation after resettlement. In: Why it matters. Utah Valley University, Orem, UT

Cairncross S, Feachem RG (1983) Environmental health engineering in the tropics: an introductory text. Wiley, Hoboken, NJ

Carbonari LT, Librelotto LI (2018) Temporary housing made from recycled paper tubes. Increase, pp 503–512. https://doi.org/10.1007/978-3-319-70272-8_39

Chaki B, Friedman A (2020) Strategies for self-help refugee shelters; learning from experiences. Hous Stud

Chaki B, Friedman A (2021) Housing the rohingya refugees in Bangladesh : learning from the experiences of self-help emergency shelters cases. J Archit Des Constr Technol 2(1):1–10

Chauhan K, Peltokorpi A, Lavikka R, Seppänen O (2019) Deciding between prefabrication and on-site construction: a choosing-by-advantage approach performance evaluation of construction companies and suppliers view project. https://doi.org/10.24928/2019/0158

Chettri N, Gautam D, Chikermane S, Prakash V, Vaghela K (2021) Sustainability assessment of Bhutanese vernacular wattle and daub houses. Innovat Infrastr Solut 6(4):1–13. https://doi.org/10.1007/S41062-021-00576-Z

Child J (2013) Beyond the minimum: a durable emergency shelter. Unpublish M. Arch Thesis, University of Waterloo, Waterloo, Canada

Clement V, Rigaud KK, de Sherbinin A, Jones B, Adamo S, Schewe J, Sadiq N, Shabahat E (2021) Groundswell part 2: acting on internal climate migration, Washington, DC

Corsellis T, Vitale A (2005) Transitional settlement: displaced populations. Oxfam-University of Cambridge, p 464

Cronin A, Shrestha D, Cornier N, Abdalla F, Ezard N, Aramburu C (2008) A review of water and sanitation provision in refugee camps in association with selected health and nutrition indicators—The need for integrated service provision. J Water Health 6:1–13. https://doi.org/10.2166/wh.2007.019

Culbertson S, Dimarogonas J, Costello K, Lanna S (2019) Applying technology to the global refugee crisis. In: RAND social and economic well-being

Cuny FC (1977) Refugee camps and camp planning: the state of the art. Disasters 1(2):125–143. https://doi.org/10.1111/j.1467-7717.1977.tb00020.x

Dabaieh M, Emami N, Heinonen JT, Marteinsson B (2020) A life cycle assessment of a 'minus carbon' refugee house: global warming potential and sensitivity analysis. Archnet-IJAR 14(3):559–579. https://doi.org/10.1108/ARCH-11-2019-0258/FULL/XML

Daher E, Kubicki S, Halin G (2015) A parametric process for shelters and refugees' camps design. eCAADe 2. https://doi.org/10.52842/conf.ecaade.2015.2.541

Dalal A, Darweesh A, Misselwitz P, Steigemann A (2018) Planning the ideal refugee camp? A critical interrogation of recent planning innovations in Jordan and Germany. Urban Plan 3(4):64–78. https://doi.org/10.17645/up.v3i4.1726

Daniel D, Jane B, Mark B (2011) Untangling parametric schemata: enhancing collaboration through modular programming. In: Leclercq P (ed) CAAD futures 2011: designing together. University of Liege, Liege

De Chiara J, Koppelman L (1975) Urban planning and design criteria, 2nd edn. Van Nostrand Reinhold, New York, NY

de Rooij LL, Wascher DM, Paulissen MPCP (2016) Sustainable design principles for refugee camps. Wageningen Environmental Research, Wageningen

De Sousa C (2021) Sustainable brownfield development: building a sustainable future on sites of our polluting past. Taylor & Francis Ltd., London

Devictor X, Do QT (2017) How many years have refugees been in exile? Popul Dev Rev 43(2):355–369. https://doi.org/10.1111/padr.12061

Duclos D, Sharma E, Banke-Thomas A, Castaner MM (2024) Unsafe and unstable housing for refugees and asylum seekers is a threat to intergenerational health. BMJ 384:q561. https://doi.org/10.1136/bmj.q561

Duncan CI (1992) Soils and foundations for architects and engineers. Springer, New York, NY

European Civil Protection and Humanitarian Aid Operations (ECHO) (2017) Humanitarian shelter and settlements

Fernando R, Drogemuller R, Burden A (2012) Parametric and generative methods with building information modelling: connecting BIM with explorative design modelling. Beyond Codes and Pixels—Proceedings of the 17th International Conference on Computer-Aided Architectural Design Research in Asia, CAADRIA 2012: pp 537–546. https://doi.org/10.52842/CONF.CAADRIA.2012.537

Fosas D, Albadra D, Natarajan S, Coley D (2018) Improving the thermal comfort in new shelters: design recommendations for Azraq refugee camp. https://doi.org/10.6084/M9.FIGSHARE.8977556

Fratzscher M, Junker S (2015) Integrating refugees: a long-term, worthwhile investment. DIW Econ Bull 3:612–616

Frey C, Gatzweiler M (2018) How tech can bring dignity to refugees in humanitarian crises. The Conversation

Friedman A (2012) Fundamentals of sustainable dwellings. Island Press/Center for Resource Economics, Washington, DC

Friedman A (2018) Smart homes and communities. Images Publishing Dist Ac, Mulgrave

Friedman A (2021) Pre-fab living. Thames & Hudson, London/New York, NY

Gehl J (2024) Life between buildings, 6th edn. Island Press, Washington, DC

Geoscience E and MD (n.d.) Refusing to disappear: how the people of Tukuraki became symbols of a Resilient Pacific. https://gem.spc.int/news/2021/06/refusing-to-disappear-how-the-people-of-tukuraki-became-symbols-of-a-resilient-pacific. Accessed 13 Aug 2021

Ghandehari H (2000) Kit homes as a self-help affordable strategy. McGill University

Ghisleni C (2021) Shelter architecture: the subjective aspects of migrant and refugee settlement projects. In: ArchDaily. https://www.archdaily.com/966887/shelter-architecture-the-subjective-aspects-of-migrant-and-refugee-settlement-projects. Accessed 18 Sep 2024

Gilmer M (2019) Living on flood plains and wetlands: a homeowner's handbook. Lyons Press

Global Shelter Cluster (2021a) Shelter projects 8th edition, chrome-extension://efaidnbmnnnibpcajpcglclefindmkaj/ https://shelterprojects.org/essentials/shelter-projects-essentials-2021-print.pdf. Accessed 30 May 2024

Global Shelter Cluster (2021b) Shelter projects essentials-learning from programming in humanitarian crises, chrome-extension://efaidnbmnnnibpcajpcglclefindmkaj/ https://shelterprojects.org/essentials/shelter-projects-essentials-2021-print.pdf. Accessed 30 May 2024

Global Shelter Clusters (2018a) Bangladesh 2017–2018/Rohingya Crisis, chrome-extension://efaidnbmnnnibpcajpcglclefindmkaj/ https://shelterprojects.org/shelterprojects2017-2018/SP17-18_A13-A15-Bangladesh-2017-2018.pdf. Accessed 30 May 2024

Global Shelter Clusters (2018b) Site planning. Shelter projects. https://sheltercluster.org/resources/documents/state-humanitarian-shelter-and-settlements-2018pdf. Accessed 30 May 2024

Graham J, Miller S (2021) How Kenya can create shared growth by facilitating economic inclusion for refugees

Grant J (2002) Mixed use in theory and practice: Canadian experience with implementing a planning principle. J Am Plann Assoc 68(1):71–84. https://doi.org/10.1080/01944360208977192

Green Building (2017) Indoor air quality. http://www.greenbuilding.com/indoor-air-quality/ . Accessed 12 Jun 2022

Gupta M (2012) Settlement and shelter reconstruction. In: Handbook of hazards and disaster risk reduction, pp 553–566. https://doi.org/10.4324/9780367854584-49/SETTLEMENT-SHELTER-RECONSTRUCTION-ERIN-CONNELL-BRENT-DOBERSTEIN

Habib RR, Basma SH, Yeretzian JS (2006) Harboring illnesses: on the association between disease and living conditions in a Palestinian refugee camp in Lebanon. Int J Environ Health Res 16(2):99–111. https://doi.org/10.1080/09603120500538341

Haganuma S, Urabe T, Ishisaka K, Wood Studio Co H (2015) A study on the reuse characteristic of wooden temporary housing—a study on wooden temporary housing in fukushima prefecture after the Tohoku earthquake. J Archit Plann, AIJ 80(710):813–822

Hall E (2007) Divide and sprawl, decline and fall: a comparative critique of Euclidean zoning. Univ Pitt Law Rev 68:915–952

Harms H (1992) Self-help housing in developed and third world countries. In: Housing policy in the socialist third worldpolicy in the socialist third world. Mansell, London

Harris R (1999) Slipping through the cracks: The origins of aided self-help housing, 1918-53. Hous Stud 14(3):281–309. https://doi.org/10.1080/02673039982803

Harrouk C (2020) Refugee camps: from temporary settlements to permanent dwellings. In: ArchDaily. https://www.archdaily.com/940384/refugee-camps-from-temporary-settlements-to-permanent-dwellings. Accessed 14 May 2021

Harvey P (2004) Excreta disposal in emergencies. (October): p 88

Hersperger AM (2006) Spatial adjacencies and interactions: neighborhood mosaics for landscape ecological planning. Landsc Urban Plan 77(3):227–239. https://doi.org/10.1016/j.landurbplan.2005.02.009

Holden R, Liversedge J (2011) Construction for landscape architecture. Laurence King Publishing

Humanitarian Library (2020) Flood resilient shelter reconstruction, retrofitting and repair technical guidelines

Humanitarian Response (2021) Emergency preparedness and response plan, chrome-extension://efaidnbmnnnibpcajpcglclefindmkaj/ https://www.un.org/unispal/wp-content/uploads/2021/12/OCHAHUMRESPLAN2022_161221.pdf. Accessed 16 May 2024

Hussain J, Khan A, Zhou K (2020) The impact of natural resource depletion on energy use and CO2 emission in Belt & Road Initiative countries: a cross-country analysis. Energy 199:117409. https://doi.org/10.1016/J.ENERGY.2020.117409

Interiano-Shiverdecker CG, Kondili E, Bazzi R (2022) Refugees in context: a systemic, intersectional exploration of discrimination in the United States. Int J Adv Counsel 44(2):263–282. https://doi.org/10.1007/s10447-021-09457-7

International Organization for Migration (2019) Thousands displaced as monsoon rains pound Rohingya camps. In: International Organization for Migration. https://www.iom.int/news/thousands-displaced-monsoon-rains-pound-rohingya-camps. Accessed 1 Mar 2024

International Organization for Migration (IOM) (2023) Shelter projects

Islam MM, Barman A, Khan MI, Goswami GG, Siddiqi B, Mukul SA (2022) Sustainable livelihood for displaced rohingyas and their resilience at Bhashan char in Bangladesh. Sustainability 14(10):6374. https://doi.org/10.3390/su14106374

JICA (Japan International Cooperation Agency) (2013) The study of reconstruction processes from large-scale disasters. Tokyo, Japan

Joglekar SN, Kharkar RA, Mandavgane SA, Kulkarni BD (2018) Sustainability assessment of brick work for low-cost housing: a comparison between waste-based bricks and burnt clay bricks. Sustain Cities Soc 37:396–406. https://doi.org/10.1016/J.SCS.2017.11.025

Kapur P (1989) From ideas to practice : "self-help" in housing from interpretation to application. Massachusets Institute of Technology (MIT), Boston, MA

Karmod Refugee Housing Unit | Asylum-seekers and Migrants (n.d.) In: Karmod. https://karmod.com/en/page/refugee-housing-unit. Accessed 10 Jul 2022

Kazerooni Y, Gyedu A, Burnham G, Nwomeh B, Charles A, Mishra B, Kuah SS, Kushner AL, Stewart BT (2016) Fires in refugee and displaced persons settlements: The current situation and opportunities to improve fire prevention and control. Burns 42(5):1036–1046. https://doi.org/10.1016/J.BURNS.2015.11.008

Keena N, Friedman A (2024) Sustainable housing in a circular economy. Routledge, UK

Kochary S (2020) Fecal Sludge Management of Duhok City And 4 Idps and refugees camps. J Univ Duhok 23(2):309–320. https://doi.org/10.26682/csjuod.2020.23.2.25

LaGro JA Jr (2007) Site analysis: a contextual approach to sustainable land planning and site design, 2nd edn. Wiley, Hoboken, NJ

Langar SZ (2020) Beyond refugee housing: 5 examples of social infrastructure for displaced people. In: ArchDaily. https://www.archdaily.com/919496/beyond-refugee-housing-5-examples-of-social-infrastructure-for-displaced-people. Accessed 18 Sep 2024

Lewis D (2012) Use of insulating concrete forms in residential housing construction: p 85

Lopez D, Froese TM (2016) Analysis of costs and benefits of panelized and modular prefabricated homes. Procedia Eng 145:1291–1297. https://doi.org/10.1016/J.PROENG.2016.04.166

Mancini MS, Galli A, Niccolucci V, Lin D, Bastianoni S, Wackernagel M, Marchettini N (2016) Ecological footprint: refining the carbon footprint calculation. Ecol Indic 61:390–403. https://doi.org/10.1016/J.ECOLIND.2015.09.040

Mcdonnell T (2018) The refugees the world barely pays attention to. https://www.npr.org/sections/goatsandsoda/2018/06/20/621782275/the-refugees-that-the-world-barely-pays-attention-to. Accessed 30 Apr 2021

McEwan A, Marchi E, Spinelli R, Brink M (2020) Past, present and future of industrial plantation forestry and implication on future timber harvesting technology. J For Res 31(2):339–351. https://doi.org/10.1007/S11676-019-01019-3/TABLES/1

McQuaid M, Ban S (2006) Shigeru ban. Phaidon Press, London, New York, NY

Melis L (2003) Parasite paradise : a manifesto for temporary architecture and flexible urbanism. NAi Publishers/SKOR ; Available in North America through D.A.P./Distributed Art Publishers, Rotterdam/New York, NY

Moore B (2017) Refugee settlements and sustainable planning. Forced Migrat Rev

Mugahed Amran YH, El-Zeadani M, Huei Lee Y, Yong Lee Y, Murali G, Feduik R (2020) Design innovation, efficiency and applications of structural insulated panels: a review. Structure 27:1358–1379. https://doi.org/10.1016/J.ISTRUC.2020.07.044

Mukhija V (2004) The contradictions in enabling private developers of affordable housing: a cautionary case from Ahmedabad, India. Urban Studies. https://doi.org/10.1080/0042098042000268438

Mullins D (2018) Achieving policy recognition for community-based housing solutions: the case of self-help housing in England. Int J Hous Policy 18(1):143–155. https://doi.org/10.1080/19491247.2017.1384692

Mullins D, Moore T (2018) Self-organised and civil society participation in housing provision. Int J Hous Policy 18(1):1–14. https://doi.org/10.1080/19491247.2018.1422320

Nakamura S (2014) Impact of slum formalization on self-help housing construction: a case of slum notification in India. Urban Stud 51(16):3420–3444. https://doi.org/10.1177/0042098013519139

Numbers MJ (1995) Sitting a house: proper sitting and design strategies enhance energy efficiency. Fine Homebuilding, pp 40–45, February/March, 1995SPC

Parnell S, Hart D (1999) Self-help housing as a flexible instrument of state control in 20th-century South Africa. Hous Stud 14(3):367–386

Pascucci E (2017) Community infrastructures: shelter, self-reliance and polymorphic borders in urban refugee governance. Territ Politic Gov 5(3):332–345. https://doi.org/10.1080/21622671.2017.1297252

Platform on Disaster Displacement (n.d.) Key definitions. In: Disaster displacement. https://disasterdisplacement.org/the-platform/key-definitions/. Accessed 18 Feb 2024

Pooley CG, European Science Foundation (1992) Housing strategies in Europe, 1880–1930. Leicester University Press

Rethinking the Future Paper Log Houses by Shigeru Ban: Recyclable Materials and Architecture (n.d.) https://www.re-thinkingthefuture.com/case-studies/a4922-paper-log-houses-by-shigeru-ban-recyclable-materials-and-architecture/. Accessed 12 Jun 2022

Riaz MH, Khitab A, Ahmed S (2019) Evaluation of sustainable clay bricks incorporating Brick Kiln Dust. J Build Eng 24:100725. https://doi.org/10.1016/J.JOBE.2019.02.017

Rigaud KK, de Sherbinin A, Jones B, Bergmann J, Clement V, Ober K, Schewe J, Adamo S, McCusker B, Heuser S, Midgley A (2018) Groundswell : preparing for internal climate migration, Washington, DC

Roberts T, Writer R (2020) Pros & Cons of prefab homes. In: Build with rise. https://www.buildwithrise.com/stories/pros-cons-prefab-homes. Accessed 10 Jul 2022

Roseland M (2000) Sustainable community development: Integrating environmental, economic, and social objectives. Prog Plann 54(2):73–132. https://doi.org/10.1016/S0305-9006(00)00003-9

Rueda MC (2018) Prefabricated self-help kit homes- a mass customized option for affordable housing. Unpublished M. Arch Report, McGill University, Montreal, Canada

Salih MM, Osofero AI, Imbabi MS (2020) Critical review of recent development in fiber reinforced adobe bricks for sustainable construction. Front Struct Civil Eng 14(4):839–854. https://doi.org/10.1007/S11709-020-0630-7

Sayato Y (1989) WHO guidelines for drinking-water quality. Eisei kagaku 35(5):307–312. https://doi.org/10.1248/jhs1956.35.307

Scott-Smith T, Breeze ME (eds) (2020) Structures of protection?: Rethinking refugee shelter, 1st edn. Berghahn Books, Oxford, NY

Self-help Housing Organization (n.d.) What is self help housing? | Self help housing. https://self-help-housing.org/what-is-self-help-housing/. Accessed 22 May 2022

Shaw M (2015) These modular homes enable plug-and-play urban preservation. Architizer J. https://architizer.com/blog/inspiration/stories/pao-courtyard-house-plugin/. Accessed 16 Jul 2022

Shelter Centre (2009) Transitional shelter guidelines. Geneva, chrome-extension://efaidnbmnnnibpcajpcglclefindmkaj/ https://www.iom.int/sites/g/files/tmzbdl486/files/migrated_files/What-We-Do/docs/Transitional-Shelter-Guidelines.pdf. Accessed 30 May 2024

Shelter Centre (2012) Transitional shelter guidelines. Geneva, chrome-extension://efaidnbmnnnibpcajpcglclefindmkaj/ https://www.iom.int/sites/g/files/tmzbdl486/files/our_work/Shelter/documents/Transitional%20Shelter%20Guidelines%20-%202012.pdf. Accessed 30 May 2024

SPC Geoscience E and MD (n.d.) Refusing to disappear: how the people of Tukuraki became symbols of a Resilient Pacific. https://gem.spc.int/news/2021/06/refusing-to-disappear-how-the-people-of-tukuraki-became-symbols-of-a-resilient-pacific. Accessed 13 Aug 2021

Sphere (2018) The sphere handbook humanitarian charter and minimum standards in humanitarian response, Switzerland, Geneva

Sundaram A (2012) Impact of self-help group in socio-economic development of India. IOSR J Hum Soc Sci 5(1):20–27

Swamy RN (2015) Holistic design: key to sustainability in concrete construction. Proc Inst Civil Eng Struct Build 146(4):371–379. https://doi.org/10.1680/STBU.2001.146.4.371

Teasdale S, Jones PA, Mullins D (2009) Briefing Paper 53 Connecting the dots : the potential for self-help housing to address homelessness. Management (January): pp 1–27

The Guardian (2017) Why Ikea's flatpack refugee shelter won design of the year | Design | The Guardian. https://www.theguardian.com/artanddesign/2017/jan/27/why-ikea-flatpack-refugee-shelter-won-design-of-the-year. Accessed 10 Jul 2022

The liveable and adaptable home | YourHome (n.d.) https://www.yourhome.gov.au/live-adapt/liveable-adaptable-home. Accessed 16 Jul 2022

The Nansen Initiative (2015) Agenda for the protection of cross-border displaced persons in the context of disasters and climate change, vol 1. https://disasterdisplacement.org/resource/nipa-vol1/. Accessed 1 June 2024

The UN Refugee Agency (2019) Camp planning standards (planned settlements). UNHCR Emerg Handbook 183(183):1–8

The UN Refugee Agency (n.d.-a) Government partners. https://www.unhcr.org/donors.html. Accessed 14 May 2021

The UN Refugee Agency (n.d.-b) What is a refugee? Definition and meaning. https://www.unrefugees.org/refugee-facts/what-is-a-refugee/. Accessed 30 Apr 2021

Tu C, Suweis S, D'Odorico P (2019) Impact of globalization on the resilience and sustainability of natural resources. Nat Sustain 2(4):283–289. https://doi.org/10.1038/s41893-019-0260-z

Turner A, Pathirana S, Daley A, Gill PS (2009) Sri Lankan tsunami refugees: a cross sectional study of the relationships between housing conditions and self-reported health. BMC Int Health Hum Rights 9(1):16. https://doi.org/10.1186/1472-698X-9-16

Tyrer RA, Fazel M (2014) School and community-based interventions for refugee and asylum seeking children: a systematic review. PloS One 9(2):e89359. https://doi.org/10.1371/JOURNAL.PONE.0089359

UNHCR (1989) Supplies and food aid- field handbook. UNHCR

UNHCR (1992) UNHCR manual-water supply in refugee situations, programme and technical support section, Geneva. https://doi.org/10.4324/9780203850688

UNHCR (2007) Handbook for emergencies united nations high commissioner for refugees, 2nd edn

UNHCR (2020) Trends at a glance: global trends forced displacement in 2019. UNHCR The UN Refugee Agency: pp 1–84

UNHCR (2021) Shelter and sustainability, chrome-extension://efaidnbmnnnibpcajpcglclefindmkaj/ https://emergency.unhcr.org/sites/default/files/2024-01/UNHCR%20shelter%20and%20sustainability%20overview%20April%202021.pdf. Accessed 31 May 2024

UNHCR (2024) Emergency shelter solutions and standards. https://emergency.unhcr.org/emergency-assistance/shelter-camp-and-settlement/shelter-and-housing/emergency-shelter-solutions-and-standards. Accessed 2 Oct 2024

UNHCR (2024a) Refugee data finder. https://www.unhcr.org/refugee-statistics/. Accessed 2 Oct 2024

UNHCR (2024b) Global trends—forced displacement in 2023

UNHCR (n.d.) Emergency shelter standard. In: UNHCR emergency handbook. https://emergency.unhcr.org/entry/36774/emergency-shelter-standard. Accessed 21 May 2022

UNHCR PPRE contingency planning (refugee situations)—UNHCR|Emergency handbook (n.d.) https://emergency.unhcr.org/entry/35282/ppre-contingency-planning-refugee-situations#1,1660794853153. Accessed 16 Aug 2022

UNHCR Preparedness Package for Refugee Emergencies (PPRE)—UNHCR|Emergency Handbook (n.d.) https://emergency.unhcr.org/entry/34912/preparedness-package-for-refugee-emergencies-ppre. Accessed 16 Aug 2022

UNHCR TURA (1997) UNHCR Commodity Distribution: A Practical Guide for Field staff. (June): p 77

UNHCR UNHCR (n.d.) The 1951 Refugee Convention. https://www.unhcr.org/1951-refugee-convention.html. Accessed 17 May 2022
UNHCR—History of UNHCR (n.d.) https://www.unhcr.org/history-of-unhcr.html. Accessed 21 Feb 2020
UNHCR—Shelter (n.d.) https://www.unhcr.org/shelter.html. Accessed 21 Feb 2020
Unite for Sight (n.d.) Food, water, sanitation, and housing in refugee camps. http://www.uniteforsight.org/refugee-health/module3#_ftn20. Accessed 30 Apr 2021
United Nations High Commissioner for Refugees (2007) Handbook for emergencies. United Nations High Commissioner for Refugees, Geneva
United Nations High Commissioner for Refugees (2015a) https://www.unhcr.org/media/unhcr-global-trends-2015. Accessed 30 May 2024
United Nations High Commissioner for Refugees (2015b) Handbook for emergencies, 4th edn, Geneva. https://emergency.unhcr.org/. Accessed 1 June 2024
Van der Lugt P, van den Dobbelsteen AAJF, Janssen JJA (2006) An environmental, economic and practical assessment of bamboo as a building material for supporting structures. Construct Build Mater 20(9):648–656. https://doi.org/10.1016/J.CONBUILDMAT.2005.02.023
Venkatarama Reddy BV (2009) Sustainable materials for low carbon buildings. Int J Low-Carbon Technol 4(3):175–181. https://doi.org/10.1093/IJLCT/CTP025
Ward PM (1983) Self-help housing: a critique. Alexandrine Press, Mansell Publishing Ltd.
Ward P (2012) Self-help housing ideas and practice in the Americas, pp 283–310
WHO-Geneva, CH, World Health Organization (1988) Urban vector and pest control: eleventh report of the WHO Expert Committee on Vector Biology and Control, Geneva
Wietek B (2022) Slopes and excavations, 1st edn. Springer, Wiesbaden
Winchester N, Reilly JM (2020) The economic and emissions benefits of engineered wood products in a low-carbon future. Energy Econ 85:104596. https://doi.org/10.1016/J.ENECO.2019.104596
Winchip SM (2011) Sustainable design for interior environments. Fairchild Books, New York, NY
Woodard AC, Milner HR (2016) Sustainability of timber and wood in construction. In: Sustainability of construction materials, pp 129–157. https://doi.org/10.1016/B978-0-08-100370-1.00007-X
World Commission on Environment and Development (WCED) (1987) Our common future. Oxford University Press, Oxford
World Health Organization (2021) Mental health and forced displacement. World Health Organization. https://www.who.int/news-room/fact-sheets/detail/mental-health-and-forced-displacement
Yerro I (2011) Greenery in architecture: vegetation in innovative european multifamily houses. VDM Verlag Dr. Müller
Zahra T, Thamboo J, Asad M (2021) Compressive strength and deformation characteristics of concrete block masonry made with different mortars, blocks and mortar beddings types. J Build Eng 38:102213. https://doi.org/10.1016/J.JOBE.2021.102213
Zea Escamilla E, Habert G (2014) Environmental impacts of bamboo-based construction materials representing global production diversity. J Clean Prod 69:117–127. https://doi.org/10.1016/J.JCLEPRO.2014.01.067
Zhao Y, Qiu J, Xing J, Sun X (2020) Recycling of quarry dust for supplementary cementitious materials in low carbon cement. Construct Build Mater 237:117608. https://doi.org/10.1016/J.CONBUILDMAT.2019.117608
Ziersch A, Walsh M, Due C (2024) Housing and health for people from refugee and asylum-seeking backgrounds: findings from an Australian qualitative longitudinal study. BMC Public Health 24(1):1138. https://doi.org/10.1186/s12889-024-18616-5

Index

A
Accessibility and refugee health risks, 33, 112
Architect's role in humanitarian architecture, 131

B
Bamboo construction, 147, 149
Ban, S., 147, 157, 160–164, 172
Building materials, 11, 17, 84, 99, 101, 136, 147, 148, 151, 158, 159, 161, 163, 164

C
Climate change, 2, 4, 6, 17–19, 30, 34, 147, 163, 164, 174–177
Climate refugees, 21, 174
Contingency planning for displacement, 62–64
Crisis response solutions, 167

D
Disaster displacement, 2, 4
Disaster recovery shelter, 144
Displaced communities, 98, 110, 126, 128, 168, 176
Displaced populations, 2, 5, 6, 10, 30, 33, 37, 40, 59, 72, 74, 93, 129, 143, 164, 177
Displacement, 1–6, 10–12, 14, 17–19, 21, 59, 60, 62, 65, 71, 86, 98, 174
Displacement causes, 59, 65
Durable shelter, 97, 98, 101, 110, 120

E
Ecological footprint, 39, 40, 147, 148, 163, 164
Ecological footprint for refugee camp, 40, 164
Economic potential of migration, 172–174, 177
Emergency housing, 17, 21, 23, 25, 27, 32, 72, 119, 131, 134, 136, 161
Emergency preparedness, 60–62, 74
Emergency response, 60–62, 65, 74, 85, 98, 112, 127
Emergency shelter planning, 59, 74, 98, 167, 177
Emergency shelters, 1, 3, 15, 19, 40, 41, 55, 59, 64, 74, 75, 97–99, 106, 110, 112, 115, 117, 119, 120, 128, 129, 131, 136, 144, 154, 160, 164, 177, 178
Environmental impact, 42, 55, 78, 90, 92, 93, 128, 147, 152, 159, 161, 163, 171, 172, 174

F
Fire safety in refugee camp, 97, 108–109, 112
Forced displacement, 6
Forced migration, 1

G
Green materials, 157–158

H
Humanitarian architecture, 131

A. Friedman, B. Chaki, *Fundamentals of Planning and Designing Sustainable Post-Disaster Shelters*, https://doi.org/10.1007/978-3-031-83317-5

I

Indoor environmental quality for refugee shelter, 97, 106–108
Infrastructure and utilities for camp planning, 77–93
Internally displaced persons (IDPs), 30

K

Kit shelter, 138, 140

L

Land use and refugee rights, 21, 22
Liina shelter case study, 132, 137
Long-term refugee needs, 6, 9, 23, 39, 47, 77, 85, 87, 92
Low-carbon shelters, 148
Low-cost refugee shelter, 71–74

M

Modular refugee housing, 10, 123, 127, 137
Modular site planning for refugee camp, 39, 40, 43

O

Open spaces integration in camp planning, 17, 39, 41, 47

P

Parametric shelter design, 59, 67–71, 74, 75
Planning and design steps, 167–171
Prefabricated shelter prototypes, 115, 119, 128
Prefabricated shelters, 112, 115–129, 137, 172
Proximity to amenities for refugees, 21, 22, 24, 36

R

Rapid shelter delivery, 59–75
Refugee camp design, 14, 70, 71, 75, 172
Refugee camp drainage systems, 41, 50, 77–79, 92
Refugee camp flood mitigation, 106, 151
Refugee camp planning, 69
Refugee camps, 2–4, 7, 8, 10–13, 16, 17, 21–24, 26, 33, 34, 36, 37, 40, 45, 48, 52, 55, 56, 65, 68–71, 74, 75, 77, 78, 83, 87, 88, 90, 92, 93, 97, 98, 123, 124, 128, 139, 147, 148, 164, 168, 172
Refugee communities, 11, 13, 14, 23, 39, 40, 79, 88, 89, 91, 93, 123–125, 128, 135, 143, 174
Refugee community facilities, 23, 77, 80, 84–90, 93, 94, 110
Refugee crisis, 1, 60, 64, 82, 97, 124, 135, 174–177
Refugee health and well-being, 39, 92, 106
Refugee housing, 11, 13, 16, 17, 25, 26, 39, 40, 42, 43, 45, 49, 56, 59, 74, 84, 85, 104, 110, 128, 131, 134–136, 148, 164, 167, 170–174, 177
Refugee shelter design, 97, 112, 117, 167
Refugee shelter forms, 97–112
Rohingya refugee camp, 7, 77, 78, 90–93, 140, 168, 172
Rohingya refugee housing, 7, 39, 46, 52–55, 77, 78, 90–93, 140, 168, 172

S

Safety and security in refugee camp, 6, 85
Sanitary sewers in refugee camp, 77, 79–82, 94
Self-help construction, 115, 128, 129, 133, 134, 137, 138, 140, 170, 171, 176
Self-help construction training, 140–141
Self-help refugee housing, 135–136
Shelter construction methods, 79, 119, 128, 164
Shelter design, 41, 67, 68, 92, 98, 106, 108, 112, 167, 172, 176, 177
Shelter design generation, 67–68
Shelter material selection, 100
Shelter typologies, 97–101, 110
Site conditions for camp design, 45
Site selection for refugee camp, 37
Social challenges of migration, 172–174
Solid waste management in refugee camp, 77, 80, 105
Structural considerations, 105–106, 111

Sustainable emergency housing, 13, 14, 171
Sustainable emergency shelters, 167, 177, 178
Sustainable refugee housing, 16, 55
Sustainable refugee housing solutions, 59
Sustainable shelter materials, 172
Sustainable site planning, 39, 55

T

Topography and location of shelter, 39, 41
Transitional shelters, 4, 15, 21, 97–101, 110–112, 132, 137, 141, 142
Tukuraki village case study, 22, 30, 34–37, 172

V

Viable solutions for refugee shelters, 171–172

W

Wastewater treatment in refugee camp, 77, 82
Water supply in refugee camp, 34, 77, 83–84, 149, 159, 168

The manufacturer's authorised representative in the EU is Springer Nature Customer Service Centre GmbH, Europaplatz 3, 69115 Heidelberg, Germany. If you have any concerns regarding our products, please contact ProductSafety@springernature.com

Printed and bound by CPI Group (UK) Ltd, Croydon, CR0 4YY
30/09/2026
02225610-0001